Lecture Notes in Mathematics 1804

Editors:
J.-M. Morel, Cachan
F. Takens, Groningen
B. Teissier, Paris

Subseries:
Fondazione C.I.M.E., Firenze
Adviser: Pietro Zecca
and
European Mathematical Society

Springer
Berlin
Heidelberg
New York
Hong Kong
London
Milan
Paris
Tokyo

I. Cherednik   Ya. Markov R. Howe G. Lusztig

Iwahori-Hecke Algebras and their Representation Theory

Lectures given at the C.I.M.E. Summer School
held in Martina Franca, Italy
June 28 - July 6, 1999

Editors: M. Welleda Baldoni
         Dan Barbasch

Fondazione
C.I.M.E.

Springer

Authors

Ivan Cherednik
Yavor Markov
Department of Mathematics
University of North Carolina
Chapel Hill, NC 27599-3250
U.S.A

E-mail: chered@math.unc.edu
markov@math.unc.edu

Roger Howe
Department of Mathematics
Yale University
New Haven, CT 06520-8283
U.S.A

E-mail: howe@math.yale.edu

George Lusztig
Department of Mathematics
M.I.T.
Cambridge, MA 02139
U.S.A

E-mail: gyuri@math.mit.edu

Editors

M. Welleda Baldoni
Department of Mathematics
University of Rome Tor Vergata
Rome 00133
Italy

E-mail: baldoni@mat.uniroma2.it

Dan Barbasch
Department of Mathematics
547 Malott Hall
Cornell University
Ithaca, NY 14853
U.S.A.

E-mail: barbasch@math.cornell.edu

Cataloging-in-Publication Data applied for

Bibliographic information published by Die Deutsche Bibliothek

Die Deutsche Bibliothek lists this publication in the Deutsche Nationalbibliografie;
detailed bibliographic data is available in the Internet at http://dnb.ddb.de

Mathematics Subject Classification (2000): 20C08, 20C11, 19L47, 47B35

ISSN 0075-8434
ISBN 3-540-00224-3 Springer-Verlag Berlin Heidelberg New York

Springer-Verlag Berlin Heidelberg New York a member of BertelsmannSpringer
Science + Business Media GmbH

http://www.springer.de

© Springer-Verlag Berlin Heidelberg 2002
Printed in Germany

Typesetting: Camera-ready TeX output by the authors

SPIN: 10903032   41/3142/du - 543210 - Printed on acid-free paper

This text consists of notes of three courses given during the CIME summer school which took place in 1999 in Martina Franca, Italy. The subject was Iwahori-Hecke algebras and their representation theory. The program consisted of several courses taught by senior faculty and some advanced lectures given by young researchers.

The scheduled courses were

G. Heckman, *Representation theory of affine Hecke algebras*
R. Howe, *Affine-like Hecke algebras and p-adic representations theory*
G. Lusztig *Representations of affine Hecke algebras*
I.Cherednik *Hankel transform via double Hecke algebra*

The specialized lectures on more advanced topics were given by T. Haines, M. Nazarov, C. Kriloff, U. Kulkharni, K. Maktouf, J. Kim and G. Papadoupoulo.

The volume contains the notes of the courses by I. Cherednik, R. Howe, and G. Lusztig. G. Heckman was not able to provide notes for his course. We give some references later in the introduction.

In the remainder of the introduction we give some background material on affine Hecke algebras and extra references that complement the notes.

Two basic problems of representation theory are to classify irreducible representations and decompose representations occuring naturally in some other context. Algebras of Iwahori-Hecke type are one of the tools and were (probably) first considered in the context of representation theory of finite groups of Lie type. For example for $G = GL(2, \mathbb{F}_q)$, consider the question of decomposing the induced module

$$I = Ind_B^G[\mathbb{1}] := \{f : G \longrightarrow \mathbb{C} \ : \ f(gb) = f(g)\}, \qquad \iota(g)f(x) := f(g^{-1}x),$$

where B is the subgroup of upper triangular matrices. One is naturally led to consider the algebra of intertwining operators of the module I. These are linear endomorphisms $T : I \longrightarrow I$ satisfying $T \circ \iota(g) = \iota(g) \circ T$ for all $g \in G$.

This algebra of endomorphisms can be identified with the algebra of B-biinvariant functions on G with multiplication structure given by convolution. In the case of $GL(2, \mathbb{F}_q)$ it is generated (over $\mathbb{C}$) by an element T satisfying the relation $T^2 = (q - 1)T + q$, and is called the finite Hecke algebra of type

A_1. Its generalization to type A_n is of independent interest to knot theory because it is a quotient of the braid group.

The Hecke algebras mentioned above play an important role in combinatorics and representation theory of $GL(n)$ and the symmetric group. An account of their role in the theory of finite groups of Lie type is detailed in [Car] and the references therein.

The above algebra is not what Hecke introduced in the context of automorphic forms. He defined certain operators $T(n)$ ($n \in \mathbb{N}$) that act on automorphic forms for congruence groups and studied their eigenvalues and relation to Dirichlet series. An introduction to this theory can be found in [Se]. When one considers automorphic forms in the adelic setting ([Ge], [Bump]) the operators $T(n)$ are very closely related to the previous example. One is led to consider the group of rational points $G(\mathbb{F})$ for a local field $\mathbb{F}$ with residual field of characteristic p and a compact open subgroup K. The operators $T(n)$ are K-biinvariant functions supported on certain cosets. This interpretation has led to a systematic study of the representation theory of groups over totally disconnected fields.

The talks of Howe gave an overview of this research area, particularly the role of these algebras. The notes are in the article of R. Howe and C. Krillof *Affine-like Hecke algebras and p-adic representation theory* The following is a brief list of the topics covered.

– Structure of p-adic groups and their associated affine Hecke algebras.
– Bruhat and Iwahori-Bruhat decompositions from a geometric perspective.
– Application of Hecke algebras in representations of the p-adic groups.

Lusztig's talks were focused on the special case when the compact open subgroup is an Iwahori subgroup. In this case very detailed knowledge of the representation theory is available due to his work partly joint with Kazhdan. His notes, G. Lusztig *Notes on affine Hecke algebras* cover the following topics:

– The affine Hecke algebra
– $\mathcal{H}$ and equivariant K-theory
– Convolution
– Subregular case

A very different field where Hecke algebras play a role is in the area of special functions. MacDonald has made various conjectures about the existence of orthogonal polynomials in several variables attached to root systems. These polynomials are related to the spherical functions of real and p-adic groups. The conjectures have a natural interpretation in the context of representations of Iwahori-Hecke algebras. Much of the work in this area *e.g.* by Cherednik, Heckman and Opdam make essential use of this structure. Heckmann's course provided an introduction to this area. We refer to the Bourbaki talks [He] and [M].

In Cherednik's notes the focus is on the advantages of the operator approach in the theory of Bessel functions and the classical Hankel transform. An account of these results can be found in the article I. Cherednik and Y. Markov, *Hankel transform via double Hecke algebra:*

- L-operator
- Hankel transform
- Dunkl operator
- Double H double prime
- Nonsymmetric eigenfunctions
- Inverse transform and Plancherel formula
- Truncated Bessel functions

We would like to thank the speakers and the participants for their scientific work which made the school a success.

The summer school brought together young researchers from many different countries with senior faculty. Funds were provided by C.I.M.E., the National Science Foundation and the European Community.

Welleda Baldoni and Dan Barbasch

References

[Bo] A. Borel *Admissible representations of a semisimple group over a local field with vectors fixed under an Iwahori subgroup,* Invent. Math. 35, 1976, pp. 233-259

[Bump] D. Bump *Automorphic forms and representations,* Cambridge Univ. Press

[Car] R. Carter *Groups of finite Lie type, conjugacy classes and complex characters,* Wiley&Sons, 1985

[Ge] S. Gelbart *Automorphic forms on adele groups,* Annals of Math. St., Princeton University Press, vol. 83, Princeton, N.J. 1975

[He] G.J. Heckman *Dunkl operators* Séminaire Bourbaki, Asterisque 245, 1997, Exp. No. 828, pp. 223-246

[M] I.G. MacDonald *Affine Hecke algebras and orthogonal polynomials* Séminaire Bourbaki, Asterisque 237 , 1996, Exp. No. 797, pp. 189–207

[Se] J-P. Serre *Cours D'Arithmetique,* Presses Universitaires de France, 1970

Table of Contents

Hankel transform via double Hecke algebra

Ivan Cherednik[1] and Yavor Markov[2]

[1] Department of Mathematics, University of North Carolina,
Chapel Hill, NC 27599 - 3250, USA *chered@math.unc.edu*
Partially supported by NSF grant DMS–9877048
[2] Department of Mathematics, University of North Carolina,
Chapel Hill, NC 27599 -- 3250, USA *markov@math.unc.edu*

This paper is a part of the course delivered by the first author at UNC in 2000. The focus is on the advantages of the operator approach in the theory of Bessel functions and the classical Hankel transform. We start from scratch. The Bessel functions were a must for quite a few generations of mathematicians but not anymore. We mainly discuss the *master formula* expressing the Hankel transform of the product of the Bessel function by the Gaussian.

By the operator approach, we mean the usage of the Dunkl operator and the $\mathcal{H}''$, *double H double prime*, the rational degeneration of the double affine Hecke algebra. This includes the transfer from the symmetric theory to the nonsymmetric one, which is the key tool of the recent development in the theory of spherical and hypergeometric functions. In the lectures, the Hankel transform was preceded by the standard Fourier transform, which is of course nonsymmetric, and the Harish-Chandra transform, which is entirely symmetric.

We followed closely the notes of the lectures not yielding to the temptation of skipping elementary calculations. We do not discuss the history and generalizations. Let us give some references. The master formula is a particular case of that from [D]. Our proof is mainly borrowed from [C1] and [C2]. The nonsymmetric Hankel transform is due to C. Dunkl (see also [O,J]). We will see that it is equivalent to the symmetric one, as well as for the master formulas (see e.g. [L], Chapter 13.4.1, formula (9)). This is a special feature of the one-dimensional setup. Generally speaking, there is an implication nonsymmetric $\Rightarrow$ symmetric, but not otherwise.

We also study the *truncated Bessel functions*, which are necessary to treat negative half-integral k, when the eigenvalue of the Gaussian with respect to the Hankel transform is infinity. They correspond to the finite-dimensional representations of the double H double prime, which are completely described in the paper. We did not find proper references but it is unlikely that these functions never appeared before. They are very good to demonstrate the operator technique.

We thank D. Kazhdan and A. Varchenko, who stimulated the paper a great deal, M. Duflo for useful discussion, and CIME for the kind invitation.

1 L-operator

We begin with the classical operator

$$\mathcal{L} = (\frac{\partial}{\partial x})^2 + \frac{2k}{x}\frac{\partial}{\partial x}.$$

Upon the conjugation:

$$\mathcal{L} = |x|^{-k}\,\mathcal{H}\,|x|^k, \quad \mathcal{H} = (\frac{\partial}{\partial x})^2 + \frac{k(1-k)}{x^2}. \tag{1}$$

Here k is a complex number. Both operators are symmetric = even.

The φ-function is introduced as follows:

$$\mathcal{L}\varphi_\lambda(x,k) = 4\lambda^2\varphi_\lambda(x,k), \quad \varphi_\lambda(x,k) = \varphi_\lambda(-x.k), \quad \varphi_\lambda(0,k) = 1. \tag{2}$$

We will mainly write $\varphi_\lambda(x)$ instead of $\varphi_\lambda(x,k)$. Since $\mathcal{L}$ is a DO of second order, the eigenvalue problem has a two-dimensional space of solutions. The even ones form a one-dimensional subspace and the normalization condition fixes φ_λ uniquely. Indeed, the operator $\mathcal{L}$ preserves the space of even functions holomorphic at 0. The φ_λ can be of course constructed explicitly, without any references to the general theory of ODE.

We look for a solution in the form $\varphi_\lambda(x,k) = f(x\lambda,k)$. Set $x\lambda = t$. The resulting ODE is

$$\frac{d^2f}{dt^2}(t) + 2k\frac{1}{t}\frac{df}{dt}(t) - 4f(t) = 0, \quad \text{a Bessel-type equation.}$$

Its even normalized solution is given by the following series

$$f(t,k) = \sum_{m=0}^{\infty} \frac{t^{2m}}{m!\,(k+1/2)\cdots(k-1/2+m)}$$

$$= \Gamma(k+\frac{1}{2}) \sum_{m=0}^{\infty} \frac{t^{2m}}{m!\Gamma(k+1/2+m)}. \tag{3}$$

So

$$f(t,k) = \Gamma(k+\frac{1}{2})t^{-k+\frac{1}{2}}J_{k-\frac{1}{2}}(2it).$$

The existence and convergence is for all $t \in \mathbf{C}$ subject to the constraint:

$$k \neq -1/2+n, \; n \in \mathbf{Z}_+. \tag{4}$$

The symmetry $\varphi_\lambda(x,k) = \varphi_x(\lambda,k)$ plays a very important role in the theory. Here it is immediate. In the multi-dimensional setup, it is a theorem.

Let us discuss other (nonsymmetric) solutions of (3) and (2). Looking for f in the form $t^\alpha(1+ct+\ldots)$ in a neighborhood of $t = 0$, we get that the coefficients of the expansion

$$f(t) = t^{1-2k} \sum_{m=0}^{\infty} c_m t^{2m} \text{ at } t = 0$$

can be readily calculated from (3) and are well-defined for all k.

The convergence is easy to control. Generally speaking, such f are neither regular nor even. To be precise, we get even functions f regular at 0 when $k = -1/2 - n$ for an integer $n \geq 0$, i.e. when (4) does not hold. These solutions cannot be normalized as above because they vanish at 0.

Note that we do not need nonsymmetric f and the corresponding $\varphi_\lambda(x) = f(x\lambda)$ in the paper. Only even normalized φ will be considered. The nonsymmetric ψ-functions discussed in the next sections are of different nature.

Lemma 1.1. *(a) Let $\mathcal{L}^\circ$ be the adjoint operator of $\mathcal{L}$ with respect to the $\mathbf{C}$-valued scalar product $\langle f, g \rangle_0 = 2 \int_0^{+\infty} f(x)g(x)dx$. Then $|x|^{-2k} \mathcal{L}^\circ |x|^{2k} = \mathcal{L}$.*
(b) Setting $\langle f, g \rangle = 2 \int_0^{+\infty} f(x)g(x)x^{2k}dx$, the $\mathcal{L}$ is self-adjoint with respect to this scalar product, i.e. $\langle \mathcal{L}(f), g \rangle = \langle f, \mathcal{L}(g) \rangle$.

Proof. First, the operator multiplication by x is self-adjoint. Second, $(\frac{\partial}{\partial x})^\circ = -\frac{\partial}{\partial x}$ via integration by parts. Finally,

$$x^{-2k} \mathcal{L}^\circ x^{2k} = x^{-2k}((\frac{d^2}{dx^2})^\circ + (\frac{\partial}{\partial x})^\circ(\frac{2k}{x}))x^{2k} = x^{-2k}((\frac{\partial}{\partial x})^2 - \frac{\partial}{\partial x}(\frac{2k}{x}))x^{2k}$$

$$= x^{-2k}(x^{2k}(\frac{\partial}{\partial x})^2 + 4kx^{2k-1}\frac{\partial}{\partial x} + 2k(2k-1)x^{2k-2}$$

$$- 2kx^{2k-1}\frac{\partial}{\partial x} - 2k(2k-1)x^{2k-2}) = (\frac{\partial}{\partial x})^2 + \frac{2k}{x}\frac{\partial}{\partial x} = \mathcal{L}. \tag{5}$$

Therefore, $\langle \mathcal{L}(f), g \rangle = 2 \int_0^\infty \mathcal{L}(f)gx^{2k}dx =$

$$2 \int_{\mathbf{R}_+} f\mathcal{L}^\circ(x^{2k}g)\,dx = 2 \int_{\mathbf{R}_+} fx^{2k}\mathcal{L}x^{-2k}(x^{2k}g)\,dx = \langle f, \mathcal{L}(g) \rangle. \tag{6}$$

Actually this calculation is not necessary if (1) is used. Indeed, $\mathcal{H}^\circ = \mathcal{H}$.

2 Hankel transform

Let us define the *symmetric Hankel transform* on the space of continuous functions f on $\mathbf{R}$ such that $\lim_{x \to \infty} f(x)e^{cx} = 0$ for any $c \in \mathbf{R}$. Provided (4),

$$(\mathbb{F}_k f)(\lambda) = \frac{2}{\Gamma(k + 1/2)} \int_0^{+\infty} \varphi_\lambda(x, k)f(x)x^{2k}dx. \tag{7}$$

The growth condition makes the transform well-defined for all $\lambda \in \mathbf{C}$, because

$$\varphi_\lambda(x, k) \sim \text{Const}(e^{2\lambda x} + e^{-2\lambda x}) \text{ at } x = \infty.$$

The latter is standard.

We switch from $\mathbb{F}$ on functions to the transform of the operators: $\mathbb{F}(A)(\mathbb{F}(f) = \mathbb{F}(A(f))$. Remark that the Hankel transform of the function is very much different from the transform of the corresponding multiplication operator. The key point of the operator technique is the following lemma.

Lemma 2.1. *Using the upper index to denote the variable (x or λ),*

$$(a) \quad \mathbb{F}(\mathcal{L}^x) = 4\lambda^2; \quad (b) \quad \mathbb{F}(4x^2) = \mathcal{L}^\lambda; \quad (c) \quad \mathbb{F}(4x\frac{\partial}{\partial x}) = -4\lambda\frac{d}{d\lambda} - 4 - 8k.$$

Proof. Claim (a) is a direct consequence of Lemma 1.1 (b) with $g(x) = \varphi_\lambda(x)$:

$$\mathbb{F}(\mathcal{L}f) = \langle \mathcal{L}f, \varphi_\lambda \rangle = \langle f, \mathcal{L}\varphi_\lambda \rangle = 4\lambda^2 \langle f, \varphi_\lambda \rangle = 4\lambda^2 \mathbb{F}(f)$$

. Claim (b) results directly from the $x \leftrightarrow \lambda$ symmetry of ϕ, namely, from the relation $\mathcal{L}^\lambda \varphi_\lambda(x) = 4x^2 \psi_\lambda(x)$. Concerning (c), there are no reasons, generally speaking, to expect any simple Fourier transforms for the operators different from $\mathcal{L}$. However in this particular case: $[\mathcal{L}^x, x^2] = 4x\frac{\partial}{\partial x} + 2 + 4k$. Appling $\mathbb{F}$ to both sides and using (a), (b), $[4\lambda^2, \mathcal{L}^\lambda/4] = \mathbb{F}(4x\frac{\partial}{\partial x}) + 2 + 4k$. Finally

$$\mathbb{F}(4x\frac{\partial}{\partial x}) = -4\lambda\frac{d}{d\lambda} - 2 - 4k - 2 - 4k = -4\lambda\frac{d}{d\lambda} - 4 - 8k$$

Note that

$$[x\frac{\partial}{\partial x}, x^2] = 2x^2, \quad [x\frac{\partial}{\partial x}, \mathcal{L}^x] = -2\mathcal{L}^x,$$

because operators $x\frac{\partial}{\partial x}, \mathcal{L}$ are homogeneous of degree 2 and -2. So $e = x^2$, $f = -\mathcal{L}^x/4$, and $h = x\frac{\partial}{\partial x} + k + 1/2 = [e, f]$ generate a representation of the Lie algebra $sl_2(\mathbf{C})$.

Theorem 2.1. *(Master Formula) Assuming that* Re $k > -\frac{1}{2}$,

$$2\int_0^\infty \varphi_\lambda(x)\varphi_\mu(x)e^{-x^2}x^{2k}\,dx = \Gamma(k + \frac{1}{2})e^{\lambda^2 + \mu^2}\varphi_\lambda(\mu), \qquad (8)$$

$$2\int_0^\infty \varphi_\lambda(x)\exp(-\frac{\mathcal{L}}{4})(f(x))e^{-x^2}x^{2k}\,dx = \Gamma(k + \frac{1}{2})e^{\lambda^2}f(\lambda),$$

provided the existence of $\exp(-\frac{\mathcal{L}}{4})(f(x))$ *and the integral in the second formula.*

Proof. The left-hand side of the first formula equals $\Gamma(k + 1/2)\mathbb{F}(e^{-x^2}\varphi_\mu(x))$. We set

$$\varphi_\mu^-(x) = e^{-x^2}\varphi_\mu(x), \quad \varphi_\mu^+(x) = e^{x^2}\varphi_\mu(x).$$

They are eigenfunctions of the operators

$$\mathcal{L}_- = e^{-x^2} \circ \mathcal{L} \circ e^{x^2}, \quad \mathcal{L}_+ = e^{x^2} \circ \mathcal{L} \circ e^{-x^2}.$$

To be more exact, $\varphi_\mu^\pm$ is a unique eigenfunction of $\mathcal{L}^\pm$ with eigenvalue 2μ, normalized by $\varphi_\mu^\pm(0) = 1$.

Express $\mathcal{L}_-$ in terms of the operators from the previous lemma.

$$e^{-x^2} (\frac{\partial}{\partial x})^2 e^{x^2} = e^{-x^2} (e^{x^2} (\frac{\partial}{\partial x})^2 + 2(2x)e^{x^2} \frac{\partial}{\partial x} + (2 + 4x^2)e^{x^2})$$

$$= (\frac{\partial}{\partial x})^2 + 4x\frac{\partial}{\partial x} + 2 + 4x^2,$$

$$e^{-x^2} \frac{2k}{x} \frac{\partial}{\partial x} e^{x^2} = e^{-x^2} (e^{x^2} \frac{2k}{x} \frac{\partial}{\partial x} + 2xe^{x^2} \frac{2k}{x}) = \frac{2k}{x} \frac{\partial}{\partial x} + 4k,$$

$$\mathcal{L}_- = e^{-x^2} ((\frac{\partial}{\partial x})^2 + \frac{2k}{x} \frac{\partial}{\partial x}) e^{x^2} = \mathcal{L} + 4x\frac{\partial}{\partial x} + 2 + 4k + 4x^2. \quad (9)$$

Analogously, $\mathcal{L}_+ = \mathcal{L} - 4x\frac{\partial}{\partial x} - 2 - 4k + 4x^2$. Now we may use Lemma 2.1:

$$\mathbb{F}(\mathcal{L}_-^x) = \mathbb{F}(\mathcal{L}^x) + \mathbb{F}(4x^2) + \mathbb{F}(4x\frac{\partial}{\partial x}) + \mathbb{F}(2 + 4k) =$$

$$= 4\lambda^2 + \mathcal{L}^\lambda - 4\lambda\frac{d}{d\lambda} - 4 - 8k + 2 + 4k = \mathcal{L}_+^\lambda. \quad (10)$$

Thus

$$L_+^\lambda(\mathbb{F}\varphi_\mu^-) = \mathbb{F}(\mathcal{L}_-^x)(\mathbb{F}\varphi_\mu^-) = \mathbb{F}(\mathcal{L}_-^x \varphi_\mu^-) = 2\mu\mathbb{F}(\varphi_\mu^-),$$

i.e. $\mathbb{F}\varphi_\mu^-$ is an eigenfunction of $\mathcal{L}_+$ with the eigenvalue 2μ. Using the uniqueness, we conclude that $\mathbb{F}(\varphi_\mu^-)(\lambda) = C(\mu)e^{\mu^2}\varphi_\mu^+(\lambda)$ for a constant $C(\mu)$. However the left-hand side of the master formula is $\lambda \leftrightarrow \mu$ symmetric as well as $e^{\mu^2}\varphi_\mu^+(\lambda) = e^{\lambda^2 + \mu^2}\varphi_\mu(\lambda)$. So $C(\mu) = C(\lambda) = C$. Setting $\lambda = 0 = \mu$, we get the desired.

The second formula follows from the first for $f(x) = \varphi_\mu(x, k)$. Move $\exp(\mu^2)$ to the left to see this. It is linear in terms of $f(x)$ and holds for finite linear combinations of φ and infinite ones provided the convergence. So it is valid for any reasonable f. We skip the detail.

3 Dunkl operator

The above proof is straightforward. One needs the self-duality of the Hankel transform and the the commutator representation for $x\,\partial/\partial x$. The self-duality holds in the general multi-dimensional theory. The second property is more special. Also our proof does not clarify why the master formula is so simple. There is a "one-line" proof of this important formula, which can be readily generalized. It involves the *Dunkl operator*:

$$\mathcal{D} = \frac{\partial}{\partial x} - \frac{k}{x}(s - 1), \quad \text{where } s \text{ is the reflection } s(f(x)) = f(-x). \quad (11)$$

The operator $\mathcal{D}$ is not local anymore, because s is a global operator apart from a neighborhood of $x = 0$. We are going to find its eigenfunctions. Generally speaking, this may create problems since we cannot use the uniquness theorems from the theory of ODE. However everything is surprisingly smooth.

Lemma 3.1. *Considering x as the multiplication operator,*

$$s \circ x = -x \circ s, \qquad s \circ \frac{\partial}{\partial x} = -\frac{\partial}{\partial x} \circ s, \tag{12}$$

(a) $\mathcal{D}^2 = \mathcal{L}$ upon the restriction to even functions,
(b) $s \circ \mathcal{D} \circ s = -\mathcal{D}$ and $\mathcal{D}^2$ fixes the space of even functions.

Proof. Indeed, $(s \circ x)(f(x)) = s(xf(x)) = -xf(-x) = (-x \circ s)(f(x))$. The $\frac{\partial}{\partial x}$ is analogous. Then

$$\mathcal{D}^2 = \left(\frac{\partial}{\partial x}\right)^2 - \frac{k}{x}(s-1)\frac{\partial}{\partial x} - \frac{\partial}{\partial x}\frac{k}{x}(s-1) + \frac{k}{x}(s-1)\frac{k}{x}(s-1)$$

$$= \left(\frac{\partial}{\partial x}\right)^2 + \frac{k}{x}\frac{\partial}{\partial x}(s+1) - \frac{\partial}{\partial x}\frac{k}{x}(s-1) + \frac{k}{x}(s-1)\frac{k}{x}(s-1). \tag{13}$$

It is simple to calculate the final formula but unnecessary. Applying (13) to symmetric (i.e. even) functions $f(x)$, the two last terms will vanish, because $(s-1)(f(x)) = f(-x) - f(x) = 0$. So $(s+1)(f(x)) = f(-x) + f(x) = 2f(x)$, and $\mathcal{D}^2|_{even} = \left(\frac{\partial}{\partial x}\right)^2 + 2\frac{k}{x}\frac{\partial}{\partial x} = \mathcal{L}$.

   Claim (b) is obvious. Indeed, $s^2 = 1$, $s\frac{\partial}{\partial x}s = -\frac{\partial}{\partial x}s^2 = -dx$, and $s(\frac{k}{x}(s-1))s = -\frac{k}{x}s(s^2 - s) = -\frac{k}{x}(s^3 - s^2) = -\frac{k}{x}(s-1)$. Thus $s \circ \mathcal{D} \circ s = -\mathcal{D}$.

   By the way, this implies that $s \circ \mathcal{D}^2 \circ s = \mathcal{D}^2$, i.e. $\mathcal{D}^2$ commutes with s. So we do not need an explicit formula for $\mathcal{D}^2|_{even}$ to see that it preserves even functions.

   Let us consider the standard scalar product $\langle f, g \rangle_0 = \int_{-\infty}^{+\infty} f(x)g(x)dx$. Here the functions are continuous **C**-valued continues on the real line **R**. One may add the complex conjugation to g but we will not do this. The scalar product is non-degenerate, so adjoint operators are well-defined. We continue to use the notation H° for the pairing $\langle f, g \rangle_0$. Let us calculate the adjoint of $\mathcal{D}$ with respect to $|x|^{2k}$.

Proposition 3.1. *Setting $\langle f, g \rangle = \int_{-\infty}^{+\infty} f(x)g(x)|x|^{2k}dx$, the Dunkl operator $\mathcal{D}$ is anti self-adjoint with respect to this scalar product, i.e. $\langle \mathcal{D}(f), g \rangle = -\langle f, \mathcal{D}(g) \rangle$. Equivalently, $|x|^{-2k}\mathcal{D}^\circ |x|^{2k} = -\mathcal{D}$.*

Proof. Recall that $x^\circ = x$ and $\left(\frac{\partial}{\partial x}\right)^\circ = -\frac{\partial}{\partial x}$, where x is considered as the multiplication operator. Then $s^\circ = s$:

$$\langle s(f), g \rangle_0 = \int_{-\infty}^{+\infty} f(-x)g(x)dx = \int_{+\infty}^{-\infty} f(t)g(-t)(-dt) = \langle f, s(g) \rangle_0$$

for $t = -x$. Hence,

$$
\begin{aligned}
|x|^{-2k}\mathcal{D}^\circ|x|^{2k} &= |x|^{-2k}(\frac{\partial}{\partial x} - \frac{k}{x}(s-1))^\circ|x|^{2k} \\
&= |x|^{-2k}((\frac{\partial}{\partial x})^\circ - (s-1)^\circ(\frac{k}{x})^\circ)|x|^{2k} \\
&= |x|^{-2k}(-\frac{\partial}{\partial x} - (s-1)\frac{k}{x})|x|^{2k} \\
&= |x|^{-2k}(-\frac{\partial}{\partial x} + \frac{k}{x}(1+s))|x|^{2k} \\
&= |x|^{-2k}|x|^{2k}(-\frac{\partial}{\partial x}) + |x|^{-2k}(-\frac{2k}{x}|x|^{2k}) + |x|^{-2k}|x|^{2k}\frac{k}{x}(1+s) \\
&= -\frac{\partial}{\partial x} + \frac{k}{x}(s-1) = -\mathcal{D}. \qquad (14)
\end{aligned}
$$

Finally,

$$
\langle \mathcal{D}(f), g \rangle = \int_{-\infty}^{+\infty} \mathcal{D}(f(x))g(x)|x|^{2k}\, dx = \int_{-\infty}^{+\infty} f(x)\mathcal{D}^\circ(|x|^{2k}g(x))dx \quad (15)
$$

$$
= \int_{-\infty}^{+\infty} f(x)|x|^{2k}(|x|^{-2k}\mathcal{D}^\circ|x|^{2k})(g(x))dx
$$

$$
= \int_{-\infty}^{+\infty} f(x)(-\mathcal{D}(g(x)))|x|^{2k}\, dx = -\langle f, \mathcal{D}(g) \rangle. \qquad \square
$$

The proposition readily gives that $|x|^{-2k}\mathcal{L}^\circ|x|^{2k} = \mathcal{L}$ on even functions f. Indeed,

$$
\langle \mathcal{L}(f), g \rangle = \langle \mathcal{D}^2(f), g \rangle = \langle f, \mathcal{D}^2(g) \rangle = \langle f, \mathcal{L}(g) \rangle,
$$

provided that g is even too. Recall that it was not difficult to check this relation directly. In the multi-dimensional theory, this calculation is more involved and the usage of the (generalized) Dunkl operators makes perfect sense.

4 Nonsymmetric eigenfunctions

Our next step will be a study of the eigenfunctions of the Dunkl operator:

$$
\mathcal{D}\psi_\lambda(x, k) = 2\lambda\psi_\lambda(x, k), \qquad \psi_\lambda(0, k) = 1. \qquad (16)
$$

We will use the shortcut notation $f^\iota(x) = s(f(x)) = f(-x)$.

Lemma 4.1. *There exists a unique solution $\psi_\lambda(x, k)$ of (16) which is analytic for all $x \in \mathbf{R}$. It is represented in the form $\psi_\lambda(x) = g(\lambda x)$. provided that $\lambda \neq 0$. Without the normalization condition, $\psi(x)$ is unique up to proportionality in $\mathbf{R}^*$ for any λ. As $\lambda = 0$, it is given by the formula $\psi_0 = C_1 + C_2 x|x|^{-2k-1}$, where $C \in \mathbf{C}$ are arbitrary constants.*

Proof. Assuming that ψ_λ is a solution of (16), let

$$\psi_\lambda^0 = \frac{1}{2}(\psi_\lambda + \psi_\lambda^\iota), \quad \psi_\lambda^1 = \frac{1}{2}(\psi_\lambda - \psi_\lambda^\iota),$$

be its even and odd parts. By Lemma 3.1 (b), $\mathcal{D}s(\psi_\lambda(x)) = -s\mathcal{D}(\psi_\lambda(x)) = -2\lambda s(\varphi_\lambda(x))$. Hence, (16) is equivalent to

$$\mathcal{D}\psi_\lambda^0 = 2\lambda\psi_\lambda^1; \qquad \psi_\lambda^0(0) = 1$$
$$\mathcal{D}\psi_\lambda^1 = 2\lambda\psi_\lambda^0 \qquad \psi_\lambda^1(0) = 0. \tag{17}$$

Furthermore, $\mathcal{D}^2\psi_\lambda^0 = 4\lambda^2\psi_\lambda^0$. Since ψ_λ^0 is even, $\mathcal{L}\psi_\lambda^0 = 4\lambda^2\psi_\lambda$ due to Lemma 3.1. Therefore ψ_λ^0 has to coincide with φ_λ from the first section. This is true for all λ. If $\lambda \neq 0$,

$$\psi_\lambda^1 = \frac{1}{2\lambda}\mathcal{D}\psi_\lambda^0 = \frac{1}{2\lambda}\left(\frac{d\psi_\lambda^0}{dx} - \frac{k}{x}(s-1)\psi_\lambda^0\right) = \frac{1}{2\lambda}\frac{d\varphi_\lambda}{dx}. \tag{18}$$

The last equality holds because $\psi_\lambda^0 = \varphi_\lambda$ is even. Finally,

$$\psi_\lambda(x) = \varphi_\lambda(x) + \frac{1}{2\lambda}\varphi_\lambda'(x) = g(\lambda x) \text{ for } g = f + \frac{1}{2}f', \tag{19}$$

where $\varphi_\lambda(x) = f(\lambda x)$, f is from (3), and f' is the derivative. It is for $\lambda \neq 0$.

Let us consider the case $\lambda = 0$. We have $\mathcal{D}\psi_\lambda^0 = 0$, $\psi_\lambda^0(0) = 1$ and $\mathcal{D}\psi_\lambda^1 = 0$, $\psi_\lambda^1(0) = 0$. Since ψ_λ^0 is even, $\mathcal{D}\psi_\lambda^0 = \frac{d\psi_\lambda^0}{dx} = 0$. Thus $\psi_\lambda^0(x) = 1$. The ψ_λ^1 is odd. So

$$\mathcal{D}\psi_\lambda^1(x) = \frac{d\psi_\lambda^1(x)}{dx} - \frac{k}{x}(s-1)\psi_\lambda^1(x) = \frac{d\psi_\lambda^1(x)}{dx} + \frac{k}{x}(\psi_\lambda^1(x) - \psi_\lambda^1(-x))$$
$$= \frac{d\psi_\lambda^1(x)}{dx} + \frac{2k}{x}\psi_\lambda^1(1).$$

Solving the resulting ODE, $\psi_\lambda^1(x) = Cx|x|^{-2k-1}$.

In this proof, we used that $\mathcal{D}f(x) = f'(x)$ on even functions and

$$\mathcal{D}f(x) = f'(x) + \frac{k}{x}(f(x) - f(-x)) = f'(x) + \frac{2k}{x}f(x) = (\frac{\partial}{\partial x} + \frac{2k}{x})f(x)$$

on odd functions. By the way, it makes obvious the coincidence of $\mathcal{L}$ with $\mathcal{D}^2$ on even f. Indeed, $\mathcal{D}^2 f(x) = (\frac{\partial}{\partial x} + \frac{2k}{x})(\frac{\partial}{\partial x}f(x))$. For odd f, it is the other way round: $\mathcal{D}^2 f(x) = \mathcal{D}(\mathcal{D}f(x)) = \frac{\partial}{\partial x}(\mathcal{D}f(x)) = \frac{\partial}{\partial x}(\frac{\partial}{\partial x} + \frac{2k}{x})(f(x))$. In particular,

$$\mathcal{D}^2\psi_\lambda^1(x) = \frac{\partial}{\partial x}(\frac{\partial}{\partial x} + \frac{2k}{x})\psi_\lambda^1(x) = ((\frac{\partial}{\partial x})^2 + \frac{\partial}{\partial x}\frac{2k}{x})\psi_\lambda^1(x).$$

Hence $\psi_\lambda^1(x)$ is also a solution of a second order differential equation. This equation is different from that for φ, but not too different. Comparing them we come to the important definition of the *shift operator*. We show the dependence of $\mathcal{L}$ on k and set $\widetilde{\mathcal{L}_k} = (\frac{\partial}{\partial x})^2 + \frac{\partial}{\partial x}\frac{2k}{x}$.

Lemma 4.2. *(a)* $x^{-1} \circ \widetilde{\mathcal{L}}_k \circ x = \mathcal{L}_{k+1}$.

(b) $\widetilde{\mathcal{L}}_k \psi_\lambda^1 = 4\lambda^2 \psi_\lambda^1$.

(c) $(x^{-1} \circ \widetilde{\mathcal{L}}_k \circ x)(x^{-1}\psi_\lambda^1) = 4\lambda^2(x^{-1}\psi_\lambda^1)$.

Proof. The first claim:

$$x^{-1} \circ (\frac{\partial}{\partial x})^2 \circ x = x^{-1}(x(\frac{\partial}{\partial x})^2 + 2\frac{\partial}{\partial x}) = (\frac{\partial}{\partial x})^2 + \frac{2}{x}\frac{\partial}{\partial x},$$

$$x^{-1} \circ \frac{\partial}{\partial x}\frac{2k}{x} \circ x = x^{-1} \circ \frac{\partial}{\partial x} \circ 2k = \frac{2k}{x}\frac{\partial}{\partial x},$$

$$x^{-1} \circ \widetilde{\mathcal{L}}_k \circ x = (\frac{\partial}{\partial x})^2 + \frac{2}{x}\frac{\partial}{\partial x} + \frac{2k}{x}\frac{\partial}{\partial x} = (\frac{\partial}{\partial x})^2 + \frac{2(k+1)}{x}dx = \mathcal{L}_{k+1}.$$

$$(20)$$

Then $\widetilde{\mathcal{L}}_k \psi_\lambda^1 = \mathcal{D}^2\psi_\lambda^1 = 4\lambda^2\psi_\lambda^1$ due to (17). Claim (c) is a combination of (a) and (b).

Proposition 4.1. *(Shift Formula)*

$$\frac{1}{x}\psi_\lambda^1(x,k) = \frac{2\lambda}{1+2k}\psi_\lambda^0(x,k+1), \ \ i.e. \ \ \ \frac{1}{x}\frac{d\varphi_\lambda}{dx}(x,k) = \frac{4\lambda^2}{1+2k}\varphi_\lambda(x,k+1).$$

$$(21)$$

Proof. Lemma 4.2 (c) implies that $x^{-1}\psi_\lambda^1(x,k) = C(\lambda,k)\varphi_\lambda(x,k+1)$, because $\varphi_\lambda(x,k+1)$ is a unique even normalized solution of (2) for $k+1$. Thanks to (18) $\psi_\lambda^1(x,k) = (2\lambda)^{-1}\frac{d\varphi_\lambda}{dx}(x,k)$. Thus $x^{-1}\varphi_\lambda'(x,k) = C(\lambda,k)\varphi_\lambda(x,k+1)$. The constant C readily results from the expantion (3) of $\varphi_\lambda(x,k)$. Explicitly:

$$0 = (\mathcal{L}_k\varphi_\lambda - 4\lambda^2\varphi_\lambda)(0) \Rightarrow$$

$$0 = (2k+1)(x^{-1}\frac{d\varphi_\lambda}{dx})(0,k) - 4\lambda^2.$$

$$(22)$$

The shift formula can be of course checked directly without ψ_λ^1, a good exercise.

5 Master formula

Let us define the *nonsymmetric Hankel transform*. We consider complex-valued $C^\infty-$ functions f on $\mathbf{R}$ such that $\lim_{x\to\infty} f(x)e^{cx} = 0$ for any $c \in \mathbf{R}$ and set

$$(\mathcal{F}f)(\lambda) = \frac{1}{\Gamma(k+1/2)} \int_{-\infty}^{+\infty} \psi_\lambda(x,k)f(x)|x|^{2k} dx$$

$$(23)$$

We assume that $\mathrm{Re}\, k > -\frac{1}{2}$ and always take $\psi_0(x,k) = 1$. Recall that the case $\lambda = 0$ is exceptional (Lemma 4.1): the dimension of the space of eigenfunctions is 2.

Let us compute the transforms of our main operators. Compare it with Lemma 2.1 . It is much more comfortable to deal with the operators of the first order. The upper index denotes the variable.

Lemma 5.1.

$$(a) \quad \mathcal{F}(\mathcal{D}^x) = -2\lambda; \qquad (b) \quad \mathcal{F}(2x) = \mathcal{D}^\lambda; \qquad (c) \quad \mathcal{F}(s^x) = s^\lambda.$$

Proof. The first formula is an immediate consequence of Proposition 3.1 (a) with $g(x) = \psi_\lambda(x)$:

$$\mathcal{F}(\mathcal{D}f) = \langle \mathcal{D}f, \psi_\lambda \rangle = -\langle f, \mathcal{D}\psi_\lambda \rangle = -2\lambda\langle f, \psi_\lambda \rangle = -2\lambda\mathcal{F}(f).$$

Claim (b) follows from the $x \leftrightarrow \lambda$ symmetry. As to (c), use that $\psi_\lambda(-x) = \psi_{-\lambda}(x)$.

Theorem 5.1. *(Nonsymmetric Master Formula)*

$$\int_{-\infty}^{\infty} \psi_\lambda(x)\psi_\mu(x)e^{-x^2}|x|^{2k}dx = \Gamma(k + \frac{1}{2})e^{\lambda^2+\mu^2}\psi_\lambda(\mu), \qquad (24)$$

$$\int_{-\infty}^{\infty} \psi_\lambda(x) \exp(-\frac{\mathcal{D}^2}{4})(f(x)) e^{-x^2}|x|^{2k}dx = \Gamma(k + \frac{1}{2})e^{\lambda^2} f(\lambda).$$

Proof. In the first formula, the left-hand side equals $\Gamma(k+1/2)\mathcal{F}(e^{-x^2}\psi_\mu(x))$. We set

$$\psi_\mu^-(x) = e^{-x^2}\psi_\mu(x), \quad \psi_\mu^+(x) = e^{x^2}\psi_\mu(x),$$

$$\mathcal{D}_- = e^{-x^2} \circ \mathcal{D} \circ e^{x^2}, \quad \mathcal{D}_+ = e^{x^2} \circ \mathcal{D} \circ e^{-x^2}.$$

The function $\psi_\mu^\pm$ is an eigenfunction of $\mathcal{D}^\pm$ with eigenvalue 2μ. The normalization fixes it uniquely with the standard reservation about $\mu = 0$.

One gets:

$$\mathcal{D}_- = e^{-x^2}(\frac{\partial}{\partial x} - \frac{k}{x}(s-1))e^{x^2} = \frac{\partial}{\partial x} + 2x - \frac{k}{x}(s-1) = \mathcal{D} + 2x.$$

Correspondingly, $\mathcal{D}_+ = \mathcal{D} - 2x$. Using Lemma 5.1 ,

$$\mathcal{F}(\mathcal{D}_-^x) = \mathcal{F}(\mathcal{D}^x) + \mathcal{F}(2x) = -2\lambda + \mathcal{D}^\lambda = \mathcal{D}_+^\lambda.$$

Therefore

$$\mathcal{D}_+^\lambda(\mathcal{F}\psi_\mu^-) = \mathcal{F}(\mathcal{D}_-^x)(\mathcal{F}\psi_\mu^-) = \mathcal{F}(\mathcal{D}_-^x\psi_\mu^-) = 2\mu\mathcal{F}(\psi_\mu^-),$$

i.e. $\mathcal{F}(\psi_\mu^-)$ is an eigenfunction of $\mathcal{D}_+$ with the eigenvalue 2μ, and $\mathcal{F}(\psi_\mu^-)(\lambda) = C(\mu)e^{\mu^2}\psi_\mu^+(\lambda)$. Using the $\lambda \leftrightarrow \mu$-symmetry, $C(\mu) = C(\lambda) = C$ and

$$C = \int_{\mathbf{R}} e^{-x^2}|x|^{2k}dx = \Gamma(k + \frac{1}{2}).$$

Cf. the proof of the symmetric master formula.

The second formula readily follows from the first provided the existence of the function $\exp(-\frac{D^2}{4})(f(x))$ and the corresponding integral. The latter function has to go to zero at $x = \infty$ faster than e^{cx} for any $c \in \mathbf{R}$.

The symmetric master theorem is of course a particular case of (24). Indeed, we may replace $\psi_\lambda(x)$ by $2\varphi_\lambda(x) = \psi_\lambda(x) + \psi_\lambda(-x) = \psi_\lambda(x) + \psi_{-\lambda}(x)$ on the left-hand side. Then

$$\psi_\lambda(\mu) \mapsto \psi_\lambda(\mu) + \psi_{-\lambda}(\mu) = \psi_\lambda(\mu) + \psi_\lambda(-\mu) = 2\varphi_\lambda(\mu)$$

on the right-hand side. We use that the factor $e^{\lambda^2+\mu^2}$ is even. Now we either repeat the same transfer for μ or simply symmetrize the integrand.

It is more surprising that the nonsymmetric theorem can be deduced from the symmetric one. It is a special feature of the one-dimensional case. Generally speaking, there is no reason to expect such an implication. This may clarify why the nonsymmetric Hankel transform and ψ were of little importance in the classical theory of Bessel functions. They could be considered as a minor technical improvement of the symmetric theory. Now we have the opposite point of view.

Let us deduce Theorem 5.1 from Theorem 2.1. We may assume that $\lambda, \mu \neq 0$. Discarding the odd summands in the integrand,

$$\int_{\mathbf{R}} \psi_\lambda \psi_\mu e^{-x^2} |x|^{2k} dx = \int_{\mathbf{R}} (\psi_\lambda^0 + \psi_\lambda^1)(\psi_\mu^0 + \psi_\mu^1) e^{-x^2} |x|^{2k} dx$$

$$= \int_{\mathbf{R}} (\psi_\lambda^0 \psi_\mu^0 + \psi_\lambda^1 \psi_\mu^1) e^{-x^2} |x|^{2k} dx = \int_{\mathbf{R}} (\varphi_\lambda \varphi_\mu + \psi_\lambda^1 \psi_\mu^1) e^{-x^2} |x|^{2k} dx. \quad (25)$$

The integral of $\varphi_\lambda \varphi_\mu$ is nothing else but (8). Let us use the shift formula to manage $\psi_\lambda^1 \psi_\mu^1$. See Proposition 4.1.

We get $(\psi_\lambda^1 \psi_\mu^1)(x, k) = \frac{4\lambda\mu x^2}{(1+2k)^2}(\varphi_\lambda \varphi_\mu)(x, k+1)$ and

$$\int_{\mathbf{R}} (\psi_\lambda^1 \psi_\mu^1)(x, k) e^{-x^2} |x|^{2k} dx = \frac{4\lambda\mu}{(1+2k)^2} \int_{\mathbf{R}} (\varphi_\lambda \varphi_\mu)(x, k+1) e^{-x^2} |x|^{2(k+1)} dx$$

$$= \frac{4\lambda\mu}{(1+2k)^2} \varphi_\lambda(\mu, k+1) e^{\lambda^2+\mu^2} \Gamma(k + \frac{3}{2})$$

$$= \frac{2\lambda\mu}{(1+2k)} \varphi_\lambda(\mu, k+1) e^{\lambda^2+\mu^2} \Gamma(k + \frac{1}{2})$$

$$= \psi_\lambda^1(\mu, k) e^{\lambda^2+\mu^2} \Gamma(k + \frac{1}{2}). \quad (26)$$

This concludes the deduction.

6 Double H double prime

Let $\mathcal{H}''$ be the double degeneration of the double affine Hecke algebra:

$$\mathcal{H}'' = \langle\, \partial,\, x,\, s \mid sxs = -x, \quad s\partial s = -\partial, \quad [\partial, x] = 1 + 2ks\,\rangle. \qquad (27)$$

Its *polynomial representation* $\rho : \mathcal{H}'' \to \mathrm{End}(\mathcal{P})$ in $\mathcal{P} = \mathbf{C}[x]$ is as follows:

$$\rho(x) = \text{ multiplication by } x, \qquad \rho(s) = s, \qquad \rho(\partial) = \mathcal{D},$$

where s is the reflection $f \mapsto f^\iota$, D is the Dunkl operator. The first two of the defining relations of $\mathcal{H}''$ are satisfied thanks to Lemma 3.1. As to the last,

$$[\mathcal{D}, x] = \left(\frac{\partial}{\partial x} - \frac{k}{x}(s-1)\right)x - x\left(\frac{\partial}{\partial x} - \frac{k}{x}(s-1)\right)$$

$$= x\frac{\partial}{\partial x} + 1 - x\frac{k}{x}(-s-1) - x\frac{\partial}{\partial x} + x\frac{k}{x}(s-1) = 1 + 2ks. \qquad (28)$$

Theorem 6.1.
(a) *Any nonzero finite linear combination* $H = \displaystyle\sum_{m,n,\epsilon} c_{m,n,\epsilon} x^m \partial^n s^\epsilon$, *where*

$n, m \in \mathbf{Z}_+$, $\epsilon \in \{0, 1\}$, *acts as a nonzero operator in* $\mathcal{P}$.

(b) *Any element* $H \in \mathcal{H}''$ *can be uniquely expressed in the form* $H = \displaystyle\sum_{m,n,\epsilon} x^m \partial^n s^\epsilon$. *The representation* ρ *is faithful for any* k.

Proof. Let $H = \displaystyle\sum_{n,\epsilon} f_n \partial^n s^\epsilon = \sum_{m=0}^{N}(g_m \partial^m)(1+s) + \sum_{m=0}^{M}(h_m \partial^m)(1-s)$ for polynomials f_m, g_m, and h_m. We may assume that and at least one of the leading coefficients $g_N(x)$, $h_M(x)$ is nonzero. Then $\rho(H) = L_+(1+s) + L_-(1-s)$ for differential operators $L_+ = g_N(x)(\frac{\partial}{\partial x})^N + \dots$ and $L_- = h_M(x)(\frac{\partial}{\partial x})^M + \dots$ modulo differential operators of lower orders. Applying $\rho(H)$ to even and odd functions, we get that $\rho(H) = 0$ implies that both L_+ and L_- have infinite dimensional spaces of eigenfunctions. This is impossible. Claim (a) is verified.

Concerning (b), any element $H \in \mathcal{H}''$ can be obviously expressed in the desired form. Such expression is unique and the representation ρ is faithful thanks to (a).

The theorem is the key point of the representation theory of the double H. It is a variant of the so-called PBW theorem. There are not many algebras in mathematics and physics possessing this property. All have important applications. The double Hecke algebra is one of them.

Next, we study the irreducibility of ρ.

Lemma 6.1. *The Dunkl operator* $\mathcal{D}$ *has only one eigenvalue in* $\mathcal{P}$, *namely,* $\lambda = 0$. *If* $k \neq -1/2 - n$ *for any* $n \in \mathbf{Z}_+$, *then* $\mathcal{D}$ *has a unique (up to a constant) eigenfunction in* $\mathcal{P}$, *the constant function* 1. *When* $k = -1/2 - n$ *for* $n \in \mathbf{Z}_+$, *the space of* 0-*eigenfunctions is* $\mathbf{C} + \mathbf{C}x^{2n+1}$.

Proof. Let $p(x) \in \mathcal{P}$ be an eigenfunction for $\mathcal{D}$. Since $\mathcal{D}$ lowers the degree of any polynomial by 1, we have $\mathcal{D}^{m+1}p = 0$, where $m = \deg p$. Therefore all eigenvalues of $\mathcal{D}$ are zero. Representing p as the sum $p(x) = p^0(x) + p^1(x)$ of even p^0 and odd p^1, $\mathcal{D}p = \frac{\partial}{\partial x}p^0 + (\frac{\partial}{\partial x}p^1 + \frac{2k}{x}p^1) = 0$. Both the even and the odd parts of this expression have to be zero. Hence $\frac{\partial}{\partial x}p^0 = 0$ and $\frac{\partial}{\partial x}p^1 + \frac{2k}{x}p^1 = 0$. Therefore $p^0 =$ Const. Setting $p^1(x) = \sum a_l x^{2l+1}$,

$$(\frac{\partial}{\partial x} + \frac{2k}{x})p^1(x) = \sum a_l(2l + 1 + 2k)x^{2l} = 0.$$

If $k \neq -1/2 - n$ for any $n \in \mathbf{Z}_+$ then $a_l = 0$ for any l, i.e $p^1 = 0$. Otherwise $k = -1/2 - n$ for a certain $n \in \mathbf{Z}_+$ and $p^1(x)$ is proportional to x^{2n+1}.

Theorem 6.2.
(a) The representation ρ is irreducible if and only if $k \neq -1/2 - n$ for any $n \in \mathbf{Z}_+$.
(b) If $k = -1/2 - n$ for $n \in \mathbf{Z}_+$, then there exists a unique non-trivial $\mathcal{H}''$ - submodule of $\mathcal{P}$, namely, $(x^{2n+1}) = x^{2n+1}\mathcal{P}$.

Proof. Let $\{0\} \neq V \subset \mathcal{P}$ be a $\mathcal{H}''$ submodule of $\mathcal{P}$. Let $0 \neq v \in V$ and $m = \deg v$. Set $\mathcal{P}^{(m)} = \{p \in \mathcal{P} \mid \deg p \leq m\}$. We have $V^{(m)} = V \cap \mathcal{P}^{(m)} \neq \{0\}$. Then $V^{(m)}$ is $\mathcal{D}$ invariant. More exactly, $\mathcal{D}(V^{(m)}) \subset V^{(m-1)}$. Thus it contains an eigenfunction v_0 of $\mathcal{D}$. If $k \neq -1/2 - n$ for $n \in \mathbf{Z}_+$ then Lemma 6.1 implies that $v_0 = 1$. So $1 \in V$ and $V \ni \rho(x^m)(1) = x^m$ for any $m \in \mathbf{N}$. This means that $V = \mathcal{P}$ and completes (a).

If $k = -1/2 - n$ for $n \in \mathbf{Z}_+$ then Lemma 6.1 states that $v_0 = c_1 + c_2 x^{2n+1} \in V$ for constants c_1, c_2. If $c_1 \neq 0$ then $s(v_0) + v_0 = 2c_1 \in V$ (the latter is s-invariant and $V = \mathcal{P}$ as above. If $v_0 = x^{2n+1} \in V$ then this argument gives that $(x^{2n+1}) = x^{2n+1}\mathcal{P} \subset V$. Moreover V cannot contain the polynomials of degree less than $2n + 1$. Otherwise we can find a 0-eigenvector of $\mathcal{D}$ in the space of such polynomials, which is impossible. Hence $V = (x^{2n+1})$.

The latter is invariant with respect to x and s. Its $\mathcal{D}$-invariance readily follows from the formula $\mathcal{D}(x^l) = (l + (1 - (-1)^l)k)x^{l-1}$ considered in the range $l \geq 2n + 1$.

We can reformulate the theorem as follows. The polynomial representation has a nontrivial (proper) $\mathcal{H}''$-quotient if and only if $k = -1/2 - n$ for $n \in \mathbf{Z}_+$. In the latter case, such quotient is unique, namely, $V_{2n+1} = \mathcal{P}/(x^{2n+1})$. Its dimension is $2n + 1$.

Note that the subspace V^0_{2n+1} of V_{2n+1} generated by even polynomials is invariant with respect to the action of $h = x\frac{\partial}{\partial x} + k + 1/2$ and $e = x^2$, $f = -\mathcal{L}/4$, satisfying the defining relations of $sl_2(\mathbf{C})$ (see Section 2). We get an irreducible representation of $sl_2(\mathbf{C})$ of dimension $n + 1$.

7 Algebraization

Let us use $\mathcal{H}''$ to formalize the previous considerations and to switch to the standard terminology of the representation theory.

(a) **Inner product.** We call a representation V of $\mathcal{H}''$ *pseudo-unitary* if it possesses a non-degenerate **C**-bilinear form (u, w) such that $(Hu, w) = (u, H^\star v)$ for $H \in \mathcal{H}''$ for the anti-involution

$$\partial^\star = -\partial, \qquad s^\star = s, \qquad x^\star = x. \tag{29}$$

By anti-involution, we mean a **C**-automorphism satisfying $(AB)^\star = B^\star A^\star$. We call such form $\star$-*invariant*. Formulas (29) are compatible with the defining relations (27) of $\mathcal{H}''$ and therefore can be extended to the whole $\mathcal{H}''$. This is straightforward. For instance,

$$[x^\star, \partial^\star] = [x, -\partial] = [\partial, x] = 1 + 2ks = 1 + 2ks^\star = [\partial, x]^\star.$$

We add "pseudo" because the pairing, generally speaking, is not supposed to be positive and the functions can be complex-valued.

The pairing $\langle f, g \rangle = \int_{\mathbf{R}} f(x)g(x)|x|^{2k} dx$ gives an example, provided the existence of the integral. Taking real-valued functions, we make this inner product positive (no "pseudo"). Assuming that the functions are C^∞, we need to examine the convergence at $x = 0$ and $x = \infty$. If $\mathrm{Re}\,(k) > -1/2$ then it suffices to take regular f at $x = 0$. At infinity, $f(x)|x|^k$ has to be of type $L^1(\mathbf{R})$. Polynomials times the Gaussian e^{-x^2} are fine.

(b) **Gaussians.** A homomorphism $\gamma : V \to W$ for two $\mathcal{H}''$-modules V, W is called a *Gaussian* if $\gamma H = \tau(H)\gamma$ for the following automorphism τ of $\mathcal{H}''$:

$$\tau(\partial) = \partial - 2x, \qquad \tau(x) = x, \qquad \tau(s) = s. \tag{30}$$

These formulas can be extended to an automorphism of $\mathcal{H}''$. Indeed, $\tau(s)\tau(\partial)\tau(s) = s(\partial + 2x)s = -\partial - 2x = -\tau(\partial)$, the same holds for x, and

$$[\tau(\partial), \tau(x)] = [d - 2x, x] = 1 + 2s = \tau(1 + 2s).$$

Note that 2 can be replaced by any constant $\alpha \in \mathbf{C}$ in this definition. We get a family of automorphisms $\tau_\alpha(\partial) = \partial - 2\alpha x$ of $\mathcal{H}''$. They lead to the following generalization of the master formula:

$$\int_{\mathbf{R}} \psi_\lambda(x)\psi_\mu(x)e^{-\alpha x^2}|x|^{2k} dx = \psi_\lambda(\frac{\mu}{\alpha})e^{\frac{\lambda^2 + \mu^2}{\alpha}}\alpha^{-k}\Gamma(k + \frac{1}{2}).$$

Here $\alpha > 0$ to ensure the convergence. The substitution $u = \sqrt{\alpha}x$ readily makes it equivalent to (24): use that $\psi_\lambda(x)$ is a function of the $x\lambda$. One can also follow the proof of the master formula employing $e^{-\alpha x^2}\mathcal{D}e^{\alpha x^2} = \mathcal{D} + 2\alpha x$.

If representations V, W are algebras of functions on the same set then γ can be assumed to be a function, to be more exact, the operator of multiplication by a function. For instance, the multiplication by e^{x^2} sends the polynomial representation $\mathcal{P}$ to $\mathcal{P}e^{\pm x^2}$. The latter is a $\mathcal{H}''$-module too. Adding all integral powers of e^{x^2} to $\mathcal{P}$ we make this multiplication an inner automorphism of the resulting algebra. However it is somewhat artificial. Algebraically, the resulting representation $\mathcal{P}[e^{mx^2}, m \in \mathbf{Z}]$ is "too" reducible. Analytically, we mix together e^{x^2} and e^{-x^2}, functions with absolutely different behaviour at infinity. There are interesting examples of inner automorphisms τ, but they are finite-dimensional.

(c) **Hankel transform.** Following Lemma 5.1, the *operator Hankel transform* is the following automorphism of $\mathcal{H}''$:

$$\sigma(s) = s, \qquad \sigma(\partial) = -2x, \qquad \sigma(2x) = \partial. \tag{31}$$

These realations are obviously comapatible with the defining relations of $\mathcal{H}''$. Any homomorphism $\mathcal{F} : V \to W$ of $\mathcal{H}''$-modules inducing σ on $\mathcal{H}''$ can be called a Hankel transform. The main example so far is $\mathcal{F} : \mathcal{P}e^{-x^2} \to \mathcal{P}e^{+x^2}$, where we identify x and λ in 23. Indeed, $\mathcal{F} \circ \mathcal{D} = -2x \circ \mathcal{F}$, $\mathcal{F} \circ 2x = \mathcal{D} \circ \mathcal{F}$, $\mathcal{F} \circ s = s \circ \mathcal{F}$ upon this identification.

It is interesting to interpret the master formula from this viewpoint. It is nothing else but the following identities for $\mathbf{F} = e^{x^2} e^{\partial^2/4} e^{x^2}$:

$$\mathbf{F}s = s\mathbf{F}, \qquad \mathbf{F}\partial = -2x\mathbf{F}, \qquad \mathbf{F}(2x) = \partial\mathbf{F}. \tag{32}$$

This means that $\mathbf{F}$ is Hankel transform whenever it is well-defined. Relations (32) can be deduced directly from the defining relations. In the first place, check that $[\partial, x^2] = 2x, [\partial^2, x] = 2\partial$. Then get that $[\partial, e^{x^2}] = 2xe^{x^2}$, $[e^{\partial^2/4}, x] = \partial e^{d^2}/2$ and use it as follows:

$$e^{x^2} e^{\partial^2/4} e^{x^2} 2x = e^{x^2} e^{\partial^2/4} (2x) e^{x^2} = e^{x^2} (\partial + 2x) e^{\partial^2} e^{x^2} =$$
$$(\partial - 2x + 2x) e^{x^2} e^{d^2} e^{x^2} = \partial e^{x^2} e^{\partial^2} e^{x^2},$$
$$e^{-x^2} e^{-\partial^2/4} e^{-x^2} 2x = e^{-x^2} e^{-\partial^2/4} (2x) e^{-x^2} = e^{-x^2} (-\partial + 2x) e^{-\partial^2} e^{-x^2} =$$
$$(-\partial - 2x + 2x) e^{x^2} e^{d^2}) e^{x^2} = -\partial e^{x^2} e^{\partial^2} e^{x^2}.$$

The commutativity of $\mathbf{F}$ with s is obvious.

Note the following *braid identity* which can be proved similarly:

$$\mathbf{F} = e^{x^2} e^{\partial^2/4} e^{x^2} = e^{\partial^2} e^{x^2/4} e^{\partial^2}.$$

Actually we do need calculations from scratch because it suffices to use the nonsymmetric master formula Lemma 5.1 and the fact that $\mathcal{P}$ is a faithful representation. For instance, $\mathbf{F}(2x)\mathbf{F}^{-1}$ and ∂ coincide in $\mathcal{P}$. The previous consideration shows that the former is an element of $\mathcal{H}''$. Hence they must coincide identically, i.e. in $\mathcal{H}''$.

A good demonstration of the convenience of such an algebraization will be the case of negative half-integral k. Before switching to this case, let us conclude the "analytic" theory calculating the inverse Hankel transform.

8 Inverse transform and Plancherel formula

Let $\operatorname{Re} k > -1/2$. We use $\psi_\lambda(x)$ from (16).

$$(\mathcal{F}_{re}f)(\lambda) = \frac{1}{\Gamma(k+1/2)} \int_{-\infty}^{+\infty} \psi_\lambda(x)f(x)|x|^{2k}\,dx, \tag{33}$$

$$(\mathcal{F}_{im}g)(x) = \frac{1}{\Gamma(k+1/2)} \int_{-i\infty}^{+i\infty} \psi_x(-\lambda)g(\lambda)|\lambda|^{2k}\,d\lambda. \tag{34}$$

The first is nothing else but $\mathcal{F}$ from (23). We just show explicitly that the integration is real.

Here we may consider **C**-valued functions f on **R** and g on **C** respectively such that

$$f(x) = o(e^{cx}) \text{ at } x = \infty\,\forall c \in \mathbf{R} \text{ and } f \in g(\lambda) = o(e^{ci\lambda}) \text{ at } \lambda = i\infty\,\forall c \in \mathbf{R}.$$

Restricting ourselves with the polynomials times the Gaussian,

$$\mathcal{F}_{re} : \mathbf{C}[x]e^{-x^2} \to \mathbf{C}[\lambda]e^{\lambda^2}, \quad \mathcal{F}_{im} : \mathbf{C}[\lambda]e^{\lambda^2} \to \mathbf{C}[x]e^{-x^2}.$$

The first map is an isomorphism. Let us discuss the latter.

Let $p(x) \in \mathbf{C}[x]$. Applying the master formula to $p_i(x) = p(ix)$,

$$\frac{1}{\Gamma(k+1/2)} \int_{-\infty}^{+\infty} \psi_\lambda(x)p(ix)e^{-x^2}|x|^{2k}\,dx = \mathcal{F}_{re}(e^{-x^2}p(ix))$$

$$= e^{\lambda^2}\exp((\mathcal{D}^\lambda)^2/4)(p(i\lambda)).$$

Since $\mathcal{D} \circ I = iI \circ \mathcal{D}$ for $I(f)(x) = f(ix)$,

$$\frac{(\mathcal{D}^\lambda)^2}{4}p(i\lambda) = \left(\frac{-(\mathcal{D}^u)^2}{4}p(u)\right)\Big|_{u=i\lambda}.$$

Now we replace $\lambda \mapsto i\lambda$, use that $\psi_{i\lambda}(x) = \psi_\lambda(ix)$, and then integrate by substitution using $z = ix$. The resulting formula reads:

$$\frac{1}{\Gamma(k+1/2)} \int_{-\infty}^{+\infty} \psi_\lambda(ix)p(ix)e^{-x^2}|x|^{2k}\,dx = e^{-\lambda^2}\exp(-(\mathcal{D}^\lambda)^2/4)(p(-\lambda))$$

$$\tag{35}$$

$$= \frac{1}{i\,\Gamma(k+1/2)} \int_{-i\infty}^{+i\infty} \psi_\lambda(z)p(z)e^{z^2}|z|^{2k}\,dz. \tag{36}$$

Switching to λ, $\mathcal{F}_{im}(e^{\lambda^2}p(\lambda)) = e^{-x^2}\exp(-(\mathcal{D}^x)^2/4)(p(x))$.

We come to the inversion theorem.

Theorem 8.1. *(Inversion Formula)* $\mathcal{F}_{im} \circ \mathcal{F}_{re} = \mathrm{id}$ *in* $\mathbf{C}[x]e^{-x^2}$, $\mathcal{F}_{re} \circ$ $\mathcal{F}_{im} = \mathrm{id}$ *in* $\mathbf{C}[\lambda]e^{\lambda^2}$.

Proof. $\mathcal{F}_{im} \circ \mathcal{F}_{re}(e^{-x^2}p(x)) = e^{-x^2}\exp(-\mathcal{D}^2/4)\exp(\mathcal{D}^2/4)(p(x)) = e^{-x^2}p(x)$. The second formula is analogous.

There is a simple "algebraic" proof based on the facts that the transform $\mathcal{F}_{im} \circ \mathcal{F}_{re}$ sends $\mathcal{D} \mapsto \mathcal{D}$, $2x \mapsto 2x$, $s \mapsto s$. Thanks to the irreducibility of $\mathbf{C}[x]e^{\pm x^2}$, we may apply the Schur lemma. However the spaces are infinite dimensional, so a minor additional consideration is necessary. We will skip it because it practically coincides with that from the proof of the Plancherel formula.

Provided $\mathrm{Re}\, k > -\frac{1}{2}$, the inner products

$$\langle f, g \rangle_{re} = \int_{-\infty}^{+\infty} f(x)g(x)|x|^{2k}dx, \qquad \langle f, g \rangle_{im} = \frac{1}{i}\int_{-i\infty}^{+i\infty} f(\lambda)g(-\lambda)|\lambda|^{2k}d\lambda \tag{37}$$

are non-degenerate respectively in $\mathbf{C}[x]e^{-x^2}$ and $\mathbf{C}[\lambda]e^{\lambda^2}$.

It is obvious when $\mathbf{R}[x], \mathbf{R}[\lambda]$ are considered instead of $\mathbf{C}[x], \mathbf{C}[\lambda]$ and $k \in \mathbf{R}$. Indeed, both forms become positive in this case. Concerning the second, use that $g(-\lambda) = \overline{g(\lambda)}$ for a real polynomial g.

For complex-valued functions and $k \in \mathbf{C}$, the claim requires proving. Let us use the irreducibility of $\mathcal{H}''-$ modules $\mathbf{C}[x]e^{-x^2}$ and $\mathbf{C}[x]e^{x^2}$, which is equivalent to the irreducibility of the polynomial representation $\mathcal{P} = \mathbf{C}[x]$, which we already know for $\mathrm{Re}\, k > -\frac{1}{2}$. Then the radical of the form $\langle \cdot, \cdot \rangle_{re}$ is a submodule of $\mathbf{C}[x]e^{-x^2}$. It does not coincide with the whole space, since

$$\langle e^{-x^2}, e^{-x^2} \rangle_{re} = \int_{\mathbf{R}} e^{-2x^2}|x|^{2k}dx = (\sqrt{2})^{-2k-1}\Gamma(k + \frac{1}{2}). \tag{38}$$

The same argument works for the imaginary integration.

Theorem 8.2. *(Plancherel Formula)*

$$\langle f, g \rangle_{re} = \langle \widehat{f}, \widehat{g} \rangle_{im} \text{ for all } f, g \in \mathbf{C}[x]e^{-x^2}, \ \widehat{f} = \mathcal{F}_{re}(f), \ \widehat{g} = \mathcal{F}_{re}(g). \tag{39}$$

Proof. Setting $\mathcal{P}_- = \mathbf{C}[x]e^{-x^2}$, we need to check that $\langle \cdot, \cdot \rangle = \langle \cdot, \cdot \rangle_{re}$ coincides with $\langle f, g \rangle_1 = \langle \widehat{f}, \widehat{g} \rangle_{im}$ for all $f, g \in \mathcal{P}_-$. In the first place, $\langle Hf, g \rangle_1 = \langle f, H^\star g \rangle_1$ for any $H \in \mathcal{H}''$, i.e. this bilinear form is $\star$-invariant. Indeed,

$$\langle \mathcal{D}f, g \rangle_1 = \langle \widehat{\mathcal{D}f}, \widehat{g} \rangle_{im} = \langle -2x\widehat{f}, \widehat{g} \rangle_{im}$$
$$= -\langle \widehat{f}, -2(-x)\widehat{g} \rangle_{im} = -\langle \widehat{f}, \widehat{\mathcal{D}g} \rangle_{im} = -\langle f, \mathcal{D}g \rangle_1.$$

Similarly, $\langle 2xf, g \rangle_1 = \langle f, 2xg \rangle_1$ and $\langle s(f), g \rangle_1 = \langle f, s(g) \rangle_1$.

Setting $\langle\,.\,,\,.\,\rangle_2 = \langle\,.\,,\,.\,\rangle - \langle\,.\,,\,.\,\rangle_1$, we get $\langle e^{-x^2}, e^{-x^2}\rangle_2 = 0$. Cf. 38. It is a $\star$-invariant form as well. Let us demonstrate that it vanishes identically.

First, $\langle e^{-x^2}, (\mathcal{D}-2x)f\rangle = \langle -(\mathcal{D}+2x)e^{-x^2}, f\rangle = \langle 0, f\rangle = 0$ for any $f \in \mathcal{P}_-$ due to the $\star$-invariance. So it is applicable to $\langle\,.\,,\,.\,\rangle_2$ too. Second, $\mathbf{C}e^{-x^2} \cap (\mathcal{D}-2x)\mathcal{P}_- = \emptyset$ because $\langle e^{-x^2}, e^{-x^2}\rangle \neq 0$. Third, $\mathcal{P}_- = \mathbf{C}e^{-x^2} \oplus (\mathcal{D}-2x)\mathcal{P}_-$. Really, the dimension of $(\mathcal{D}-2x)\mathcal{P}_{n-}$ for $\mathcal{P}_n = \mathbf{C}[1, x, \ldots, x^n]e^{-x^2}$ is $n+1$ since the kernel of the operator $\mathcal{D}-2x$ in $\mathcal{P}_-$ is zero. However $(\mathcal{D}-2x)\mathcal{P}_{n-} \subset \mathcal{P}_{(n+1)-}$. Therefore $(\mathcal{D}-2x)\mathcal{P}_{n-} = \mathcal{P}_{(n+1)-}$. Finally,

$$\langle e^{-x^2}, \mathcal{P}_-\rangle_2 = \langle e^{-x^2}, e^{-x^2} + (\mathcal{D}-2x)\mathcal{P}_-\rangle_2 = 0$$

and e^{-x^2} belongs to the radical of $\langle\,.\,,\,.\,\rangle_2$. Since the module $\mathcal{P}_-$ is irreducible, the radical has to coincide with the whole $\mathcal{P}_-$.

Taking real k, the forms $\langle\,.\,,\,.\,\rangle_{re}$ $\langle\,.\,,\,.\,\rangle_{im}$ are positive on $\mathbf{R}[x]e^{-x^2}$ and $\mathbf{R}[\lambda]e^{\lambda^2}$. The Plancherel formula allows us to complete the function spaces extending the Fourier transforms $\mathcal{F}_{re}, \mathcal{F}_{im}$ to the spaces of square integrable real-valued functions with respect to the "Bessel measure":

$$\mathcal{L}^2(\mathbf{R}, |x|^{2k}dx) \to \mathcal{L}^2(i\mathbf{R}, |\lambda|^{2k}d\lambda) \to \mathcal{L}^2(\mathbf{R}, |x|^{2k}dx).$$

The inversion and Plancherel formulas remain valid.

Here we assume that $k > -\frac{1}{2}$. Let us discuss the case of negative half-integers.

9 Finite-dimensional case

Let $k = -n - \frac{1}{2}$ for $n \in \mathbf{Z}_+$. Then $V_{2n+1} = \mathcal{P}/(x^{2n+1})$ is an irreducible representation of $\mathcal{H}''$. The elements of V_{2n+1} can be identified with polynomials of degree less than $2n + 1$.

Theorem 9.1. *Finite-dimensional representations of $\mathcal{H}''$ exist only as $k = -n - 1/2$ for $n \in \mathbf{Z}_+$. Given such k, the algebra $\mathcal{H}''$ has a unique finite-dimensional irreducible representation up to isomorphisms, namely, V_{2n+1}.*

Proof. We will use that

$$[h, x] = x, \quad [h, \partial] = -\partial \quad \text{for } h = (x\partial + \partial x)/2. \tag{40}$$

It readily follows from the defining relations of $\mathcal{H}''$. Actually (40) determines a super Lie algebra, which is $osp(2, 1)$. One may use a general theory of this algebras. For our purpose, a reduction to sl_2 is sufficient.

Note that h is $x\frac{\partial}{\partial x} + k + 1/2$ in the polynomial representation. Since the latter is faithful, (40) is exactly the claim that $x, \mathcal{D}$ are homogeneous operators of degree ± 1, which is obvious.

We will employ that $e = x^2$, $f = -\partial^2/4$, and h satisfy the defining relations of $sl_2(\mathbf{C})$. Namely, $[e, f] = h$ because

$$[\partial^2, x^2] = [\partial^2, x]x + x[\partial^2, x] = 2\partial x + x(2\partial),$$

and the relations $[h, e] = 2e, [h, f] = -2f$ readily result from (40). Cf. Section 6.

Let V be an irreducible finite-dimensional representation of $\mathcal{H}\ell''$. Then the subspaces V^0, V^1 of V formed respectively by s-invariant and s-anti-invariant vectors are preserved by h, e, and f. So they are $sl_2(\mathbf{C})$-modules. One gets

$$\partial x = h + k + 1/2, \ x\partial = h - k - 1/2 \ \text{in} \ V^0,$$

and the other way round in V^1.

Let us check that $k \in -1/2 - \mathbf{Z}_+$. All h-eigenvalues in V are integers thanks to the general theory of finite-dimensional representations of $sl_2(\mathbf{C})$. We pick a nonzero h-eigenvector $v \in V$ with the maximal possible eigenvalue m. Then $m \in \mathbf{Z}_+$ (the theory of sl_2) and $x(v) = 0$ because the latter is an h-eigenvector with the eigenvalue $m + 1$. Hence $\partial x(v) = 0$, $m + k + 1/2 = 0$, and $k = -1/2 - m$.

Let U^0 be a nonzero irreducible $sl_2(\mathbf{C})$−submodule of V^0. The spectrum of h in U^0 is $\{-n, -n+2, \ldots, n-2, n\}$ for an integer $n \geq 0$. Let $v_l \neq 0$ be an h-eigenvector with the eigenvalue l. If $e(v) = 0$ then $v = cv_n$ for a constant c, and if $f(v) = 0$ then $v = cv_{-n}$.

Let us check that $\partial x(v_n) = 0$, $x\partial(v_{-n}) = 0$, and

$$\partial x(v_l) \neq 0 \ \text{for} \ l \neq n, \ \ x\partial(v_l) \neq 0 \ \text{for} \ l \neq -n.$$

Both operators, ∂x and $x\partial$, obviously preserve U^0 :

$$\partial x(v_l) = (l + k + 1/2)v_l, \ x\partial(v_l) = (l - k - 1/2)v_l.$$

Hence,

$$\partial^2 x^2(v_l) = ((\partial x)^2 + (1 - 2k)(\partial x))(v_l) = (l + k + 1/2)(l - k + 3/2)v_l.$$

Setting $l = n$, we get that $(n + k + 1/2)(n - k + 3/2) = 0$ and $k = -1/2 - n$, because $k < 0$ and $n - k + 3/2 > 0$. Thus $\partial x(v_n) = 0$. The case of $x\partial$ is analogous.

The next claim is that $x(v_n) = 0$, $\partial(v_{-n}) = 0$. Indeed, $x(v') = 0$ and $\partial(v') = 0$ for $v' = x(v_n)$. Therefore

$$0 = [\partial, x](v') = (1 + 2ks)(v') = (1 - 1 + n)v' = nv'.$$

This means that either $v' = 0$ or $n = 0$. In the latter case, v' is proportional to v_0 and therefore $v' = x(v_0) = 0$ as well. Similarly, $\partial(v_{-n}) = 0$.

Now we use the formula

$$\partial(x^2(v_l)) = x(2 + x\partial)(v_l) = (2 + l - k - 1/2)x(v_l) = (2 + l + n)x(v_l),$$

and get that $x(v_l) \in \partial(U^0)$ for any $-n \leq l \leq n$. Hence $U = U^0 + \partial(U^0)$ is x-invariant. It is obviously ∂-invariant and s-invariant ($\Leftarrow \partial(V^0) \subset V^1$). Also the sum is direct.

Finally, U is a $\mathcal{H}''$-module and has to coincide with V because the latter was assumed to be irreducible. The above formulas are sufficient to establish a $\mathcal{H}''$- isomorphism $U \simeq V_{2n+1}$. Explicitly, the h-eigenvectors $x^i(v_{-n}) \in U$ will be identified with the monomials $x^i \in V_{2n+1}$.

Let us discuss the Hankel transform and related structures in the case of V_{2n+1}. We follow Section 7.

(a) **Form.** To make $\star$ "inner" we have to construct a non-degenerate bilinear pairing (u, v) on V_{2n+1} such that $(Hu, v) = (u, H^\star v)$. Here it is:

$$\forall f, g \in V_{2n+1} \text{ set } (f, g) = \mathrm{Res}(f(x)g(x)x^{-2n-1}), \text{ where } \mathrm{Res}(\sum a_i x^i) = a_{-1}. \tag{41}$$

The pairing is non-degenerate, because if $f = ax^l +$ lower order terms, where $a \neq 0$ and $0 \leq l \leq 2n$, then $(f, x^{2n-l}) = a$.

We introduce a scalar product $(f, g)_0 = \mathrm{Res}(fg)$ for polynomials in terms of x and x^{-1}. Denoting the conjugate of an operator A with respect to this pairing by A°,

$$s^\circ = -s, \quad x^\circ = x, \quad \frac{\partial}{\partial x}^\circ = -\frac{\partial}{\partial x}, \quad \text{and } x^{2n+1}\mathcal{D}^\circ x^{-2n-1} = -\mathcal{D}.$$

The relation $x^\circ = x$ is obvious. Concerning $\frac{\partial}{\partial x}^\circ = -\frac{\partial}{\partial x}$, it follows from the property $\mathrm{Res}(df/dx) = 0$ for any polynomial $f(x, x^{-1})$. The formula $s^\circ = -s$ results from $\mathrm{Res}(s(f(x)) = -\mathrm{Res}(f(x))$.

Switching from $^\circ$ to *, we have

$$(sf, g) = \mathrm{Res}(s(f)gx^{-2n-1}) = -\mathrm{Res}(fs(g)s(x^{-2n-1}))$$
$$= -(-1)^{2n+1}(fs(g)x^{-2n-1}) = (f, s(g)).$$

Finally,

$$\mathcal{D}^\circ = (\frac{\partial}{\partial x} + \frac{k}{x}(1 - s))^\circ = -\frac{\partial}{\partial x} + (1 + s)\frac{k}{x},$$
$$x^{-2k}\mathcal{D}^\circ x^{2k} = -\frac{\partial}{\partial x} - \frac{2k}{x} + \frac{k}{x}(1 + s) = -\frac{\partial}{\partial x} + \frac{k}{x}(-1 + s) = -\mathcal{D}. \tag{42}$$

The first equality on the second line holds because $\frac{k}{x}x^{2k} = kx^{-2n-2}$ is an even function, and thus it commutes with the action of s. Finally

$$(\mathcal{D}f, g) = (\mathcal{D}f, x^{2k}g)_0 = (f, x^{2k}x^{-2k}\mathcal{D}^\circ x^{2k}(g))_0 = -(f, x^{2k}\mathcal{D}g)_0 = -(f, \mathcal{D}g).$$

(b) **Gaussian.** The Gaussian does not exist in polynomials but of course can be introduced as a power series $e^{x^2} = \sum_{m=0}^{\infty}(x^2)^m/m!$ in the algebra of formal series $\mathbf{C}[[x]]$, a completion of the polynomial representation. The conjugation by this series induces τ on $\mathcal{H}l''$. Its inverse is $e^{-x^2} = \sum_{m=0}^{\infty}(-x^2)^m/m!$. The multiplication by the Gaussian does not preserve the space of polynomials but is well-defined on V_{2n+1} because $\forall f \in V_{2n+1}$ we have $x^m f = 0$ for $m \geq 2n + 1$. Finally,

$$\gamma^{\pm} = \sum_{m=0}^{2n}(\pm x^2)^m/m!.$$

(c) **Hankel transform.** The operator $\mathcal{D}$ is nilpotent in V_{2n+1} because it lowers the degree of $f \in V_{2n+1}$ by one. Therefore the operators $\exp(\pm\mathcal{D}^2/4) \in \mathbf{C}[[\mathcal{D}]]$ are well-defined in this representation as well as the Gaussians. It suffices to take $\sum_{m=0}^{2n}(\pm(D/2)^2)^m/m!$. Thus we may set

$$\mathbf{F} = e^{x^2}e^{\frac{\mathcal{D}^2}{4}}e^{x^2} \quad \text{in } V_{2n+1}. \tag{43}$$

Proposition 9.1. *The map $\mathcal{F}$ is a Hankel transform on V_{2n+1}, i.e. $\mathbf{F} \circ \mathcal{D} = -\mathbf{F} \circ 2x$, $\mathbf{F} \circ 2x = \mathcal{D} \circ \mathbf{F}$, $\mathbf{F}s = s\mathbf{F}$. These relations fix it uniquely up to proportionality.*

Proof. We already know that $\mathbf{F}$ is a Hankel transform (the previous section). If $\widetilde{\mathbf{F}}$ is another one then the ratio $\widetilde{\mathbf{F}}\mathbf{F}^{-1}$ commutes with $x, \mathcal{D}$, and s because of the very definition. Since V_{2n+1} is irreducible (and finite dimensional) we get that $\widetilde{\mathbf{F}}$ is proportional to $\mathbf{F}$.

10 Truncated Bessel functions

Recall that $\mathcal{D}$ has only one eigenvalue in V_{2n+1}, namely, 0. Therefore we cannot define the ψ_λ as an eigenfunction of $\mathcal{D}$ in V_{2n+1} any longer. Instead, it will be introduced as the kernel of the Hankel transform.

Any linear operator $A : V_{2n+1}^x \to V_{2n+1}^\lambda$ (the upper index indicates the variable) is a matrix. It means that

$$A(f)(\lambda) = (f(x), \alpha(x, \lambda)) = \mathrm{Res}(f(x)\alpha(x, \lambda)x^{-2n-1}), \quad \text{where}$$

$$\alpha(x, \lambda) = \sum_{l,m=0}^{2n} c_{l,m}x^l\lambda^m = \sum_{l=0}^{2n} x^{2n-l}A(x^l). \tag{44}$$

So here the kernel $\alpha(x, \alpha)$ is uniquely defined by A and vice versa.

The *truncated ψ-function* is the kernel of $\mathbf{F}$:

$$\mathbf{F}(f)(\lambda) = (f(x), \psi_\lambda(x)) = \mathrm{Res}(f(x)\psi_\lambda(x)x^{-2n-1}). \tag{45}$$

There is a somewhat different approach. Let us use that the relations from Lemma 9.1 determine $\mathbf{F}$ uniquely up to proportionality. These relations are equivalent to the following properties of $\psi_\lambda(x)$:

$$\mathcal{D}\psi_\lambda(x) = 2\lambda\psi_\lambda(x) \quad \mathrm{mod}\ (x^{2n+1}, \lambda^{2n+1}),$$
$$\psi_\lambda(x) = \psi_x(\lambda), \qquad \psi_\lambda(s(x)) = \psi_{s(\lambda)}(x). \tag{46}$$

Let us solve the first equation. Setting $\psi_\lambda(x) = \sum_{l,m=0}^{2n} c_{l,m} x^l \lambda^m$,

$$\sum_{l=1,m=0}^{l=2n,m=2n} c_{l,m} (l + (1 - (-1)^l)(-n - \frac{1}{2}))x^{l-1}\lambda^m$$

$$= \sum_{l=0,m=0}^{l=2n,m=2n-1} 2c_{l,m} x^l \lambda^{m+1} \quad \mathrm{mod}\ (x^{2n+1}, \lambda^{2n+1}),$$

$$c_{l,m} = \frac{2}{l + (1 - (-1)^l)(-n - 1/2)} c_{l-1,m-1} \text{ for } 2n \geq l > 0,\ 2n \geq m > 0,$$

where $c_{l,0} = 0 = c_{2n,m}$ for $l > 0$, $m < 2n$. $\tag{47}$

Using the $x \leftrightarrow \lambda$ symmetry, we conclude that $c_{l,0} = 0 = c_{0,l}$ for nonzero l and $c_{l,m} = 0$ for $l \neq m$. Thus

$$\psi_\lambda(x) = g_n(\lambda x) \text{ for } g_n = \sum_{l=0}^{2n} c_l t^l,\ c_l = c_{l,l},$$

where the coefficients are given by (47).

Finally, $g_n(t) = f_n(t) + (1/2)df_n/dt$ for the *truncated Bessel function* $f_n(t) = \sum_{m=0}^{n} c_{2m} t^{2m}$ which is an even solution of the truncated Bessel equation (cf. Section 1):

$$\frac{d^2 f}{dt^2}(t) + 2k\frac{1}{t}\frac{df}{dt}(t) - 4f(t) = 0 \quad \mathrm{mod}\ (t^{2n}),\ k = -n - 1/2. \tag{48}$$

This equation is sufficient to determine the coefficients of f_n uniquely for any constant term $c_0 = c_{0,0}$. They are given by the same formula (3) till c_{2n} up to proportionality. This can be checked directly using explicit formulas which will be discussed next.

Still c_0 remains arbitrary. Recall that ψ was initially introduced as the kernel of $\mathbf{F}$. So it comes with its own normalization. Let us calculate its c_0. One gets:

$$\mathbf{F}(e^{-x^2}) = e^{\lambda^2} \exp(\mathcal{D}^2/4)(e^{\lambda^2} e^{-\lambda^2}) = e^{\lambda^2} \exp(\mathcal{D}^2/4)(1) = e^{\lambda^2}, \text{ so}$$

$$\mathbf{F}(1 - \frac{x^2}{1!} + \frac{x^4}{2!} + \cdots + (-1)^n \frac{x^{2n}}{n!}) = 1 + \frac{\lambda^2}{1!} + \cdots \frac{\lambda^{2n}}{n!}. \tag{49}$$

Here the transform of 1 is proportional to λ^{2n} since the latter has to be an eigenfunction of λ, i.e. a solution of the equation $\lambda\mathbf{F}(1) = 0$ in V_{2n+1}^λ. Similarly, $\mathbf{F}(x^l) = (\mathcal{D}^\lambda/2)^l\mathbf{F}(1)$ is proportional to λ^{2n-l} for $0 \le l \le 2n$. Thus (49) leads to the relations

$$\mathbf{F}(x^{2m}) = (-1)^m \frac{m!}{(n-m)!}\lambda^{2n-2m}. \tag{50}$$

For instance, $\mathbf{F}(x^{2n}) = (-1)^n n!$. This is exactly the coefficient c_0 above.

We obtain that the normalization serving the truncated Hankel transform is

$$\psi_\lambda(0) = -n!, \; c_0 = g_n(0) = f_n(0) = -n!. \tag{51}$$

Formula (50) also results in

$$\mathbf{F}(x^{2m+1}) = \mathbf{F}(x(x^{2m})) = (\mathcal{D}/2)\mathbf{F}(x^{2m}) = (-1)^m\frac{m!}{(n-m-1)!}\lambda^{2n-2m-1}. \tag{52}$$

Substituting,

$$\psi_\lambda(x) = \sum_{m=0}^{n} x^{2n-2m}\mathbf{F}(x^{2m}) + \sum_{m=0}^{n-1} x^{2n-2m-1}\mathbf{F}(x^{2m+1}) = f_n(x\lambda) + \frac{1}{2}f_n'(x\lambda)$$

$$\text{for } f_n(t) = \sum_{m=0}^{n} \frac{(-1)^m m!}{(n-m)!}t^{2n-2m} = \sum_{m=0}^{n} \frac{(-1)^{n-m}(n-m)!}{m!}t^{2m}. \tag{53}$$

It is exactly the solution of (48) with the *truncated normalization* $f_n(0) = (-1)^n n!$.

Truncated inversion. Concluding the consideration of the case $k = -n - \frac{1}{2}$ for $n \in \mathbf{Z}_+$, let us discuss the inversion. We have the following transformations and scalar products:

$$\begin{aligned}
\mathbf{F}_+ &: \mathbf{C}[x]/(x^{2n+1}) \to \mathbf{C}[\lambda]/(\lambda^{2n+1}), & \mathbf{F}_+(f) &= \mathrm{Res}(f(x)\psi_\lambda(x)x^{-2n-1}), \\
\mathbf{F}_- &: \mathbf{C}[\lambda]/(\lambda^{2n+1}) \to \mathbf{C}[x]/(x^{2n+1}), & \mathbf{F}_-(f) &= \mathrm{Res}(f(\lambda)\psi_x(-\lambda)\lambda^{-2n-1}). \\
\langle f, g\rangle_+ &= \mathrm{Res}(f(x)g(x)x^{-2n-1}), & f, g &\in \mathbf{C}[x]/(x^{2n+1}); \\
\langle f, g\rangle_- &= \mathrm{Res}(f(-\lambda)g(\lambda)\lambda^{-2n-1}), & f, g &\in \mathbf{C}[\lambda]/(\lambda^{2n+1}).
\end{aligned} \tag{54}$$

Here $\mathbf{F}_+(f) = \mathbf{F}(f) = \hat{f}$ in the notation above. The transform $\mathbf{F}_-(f)$ coincides with $\mathbf{F}_+^\lambda(f)$ for even $f(\lambda)$ and with $-\mathbf{F}_+^\lambda(f)$ for odd $f(\lambda)$.

We can follow the "analytic" case and check that $\mathbf{F}_- \circ \mathbf{F}_+$ commutes with $\mathcal{D}, x, s$. Hence it is the multiplication by a constant thanks to the irreducibility of V_{2n+1}. The constant is $\mathbf{F}_- \circ \mathbf{F}_+(1)$ and can be readily calculated. It is equally simple to calculate all $\mathbf{F}_- \circ \mathbf{F}_+(x^l)$ using (50) and (52). For instance,

$$\mathbf{F}_- \circ \mathbf{F}_+(x^{2m}) = \mathbf{F}_-\left(\frac{(-1)^m m!}{(n-m)!}\lambda^{2n-2m}\right) =$$

$$= \frac{(-1)^m m!}{(n-m)!}\frac{(-1)^{n-m}(n-m)!}{m!}x^{2m} = (-1)^n x^{2m}.$$

Thus the truncated inversion reads:

$$\mathbf{F}_- \circ \mathbf{F}_+ = (-1)^n \, \mathrm{id} = \mathbf{F}_+ \circ \mathbf{F}_-.$$

Concerning the Plancherel formula, we may use the proportionality of the forms $\langle f,g\rangle_+$ and $\langle \widehat{f},\widehat{g}\rangle_-$ for $f,g \in V_{2n+1}$ and their transforms $\widehat{f} = \mathbf{F}(f), \widehat{g} = \mathbf{F}(g)$. It results from the irreducibility of V_{2n+1}. A direct calculation is simple as well. Let

$$\langle f,f\rangle_+ = \langle f,f\rangle = \sum_{l=0}^{2n} a_l a_{2n-l} \text{ for } f = \sum_{l=0}^{2n} a_l x^l,$$

$$\langle g,g\rangle_- = \sum_{l=0}^{2n} (-1)^l b_l b_{2n-l} \text{ for } g = \mathbf{F}(f) = \sum_{l=0}^{2n} b_l \lambda^l. \tag{55}$$

It is easy to check that

$$b_l b_{2n-l} = (-1)^{l+n} a_l a_{2n-1}.$$

Indeed, using (50) and (52):

$$b_{2m} b_{2n-2m} = (-1)^m a_{2m}\frac{m!}{(n-m)!}(-1)^{n-m} a_{2n-2m}\frac{(n-m)!}{m!}$$

$$= (-1)^n a_{2m} a_{2n-2m},$$

$$b_{2m+1} b_{2n-2m-1} =$$

$$= (-1)^m a_{2m+1}\frac{m!}{(n-m-1)!}(-1)^{n-m-1} a_{2n-2m-1}\frac{(n-m-1)!}{m!}$$

$$= (-1)^{n-1} a_{2m+1} a_{2n-2m-1}.$$

We get the truncated Plancherel formula:

$$\langle \widehat{f},\widehat{g}\rangle_- = (-1)^n \langle f,g\rangle_+.$$

The above consideration proves the coincidence for $f = g$, i.e. for the corresponding quadratic forms. It is of course sufficient.

References

[C1]   Cherednik, I.: Difference Macdonald-Mehta conjectures. IMRN **10**, 449–467 (1997).

[C2]   Cherednik, I.: One-dimensional double Hecke algebras and Gaussians, CIME (2000).

[D]   Dunkl, C.F.: Differential-difference operators associated to reflection groups, Trans. AMS. **311**, 167–183 (1989).

[J]   Jeu, M.F.E. de: The Dunkl transform, Invent. Math. **113**, 147–162 (1993).

[L]   Luke, J.: Integrals of Bessel functions, McGraw-Hill Book Company, New York-Toronto-London (1962).

[O]   Opdam, E.M.: Dunkl operators, Bessel functions and the discriminant of a finite Coxeter group, Comp. Math. **85**, 333–373 (1993).

Affine-like Hecke algebras and p-adic representation theory

Roger Howe
(Lecture Notes by Cathy Kriloff)

Department of Mathematics
Yale University
New Haven, CT 06520-8283 *howe@math.yale.edu*

These lectures begin with a discussion of the structure of p-adic groups (concentrating on GL_n and Sp_{2n}) and their associated affine Hecke algebras. The Bruhat and Iwahori-Bruhat decompositions are presented from a geometric perspective. Fundamental results and techniques from the representation theory of p-adic groups, stated without proof, are used to show the application of Hecke algebras in describing certain representations of the p-adic groups. The later part of the notes indicates why it is reasonable to believe that representations of Hecke algebras will in fact account for all representations of p-adic groups.

1 Introduction

Let G be a locally compact totally disconnected group and K be any compact open subgroup. A representation (ρ, V), consisting of a complex vector space, V and a homomorphism $\rho : G \to \mathrm{GL}(V)$, is *smooth* if the representation space V of G is a union of its K-fixed vectors, written as $V = \bigcup_K V^K$. The subgroups K form a neighborhood basis of the identity element of G and as K gets smaller, V^K gets bigger. If $\dim V^K$ is finite for all K, then the representation with space V is *admissible*. The goal is to understand unitary irreducible smooth admissible representations of G using K.

More specifically, denote by $C_c^\infty(G)$ the locally constant complex-valued functions on G with compact support. This forms an algebra under convolution and its representations are equivalent to the smooth representations of G. The *Hecke algebra of G with respect to K*, denoted $\mathcal{H}(K\backslash G/K)$ or $\mathcal{H}(G//K)$, is the subalgebra of $C_c^\infty(G)$ consisting of K bi-invariant functions on G:

$$\mathcal{H}(G//K) = \{f \in C_c^\infty(G) : f(kgk') = f(g) \text{ for all } g \in G, k, k' \in K\}.$$

This could also be written as

$$C_c^\infty(K\backslash G/K) = \left\{\sum a_g \chi_{KgK} : a_g \in \mathbb{C}\right\},$$

where χ_{KgK} is the characteristic function of the double coset KgK. Studying irreducible representations of $\mathcal{H}(G//K)$ allows us to understand irreducible representations of G with a K-fixed vector.

To explain this further, let us first explore the relationship between G and K. Assume G is *unimodular* (i.e. left and right Haar measure are equal). Normalize Haar measure so that $\mu_G(K) = \int_K dg = 1$. Then the characteristic function of K, χ_K, is idempotent in $C_c^\infty(G)$ and $\mathcal{H}(G//K) = \chi_K * C_c^\infty(G) * \chi_K$.

Given this structure of $\mathcal{H}(G//K)$, we consider the following more general context. Let A be an associative algebra over $\mathbb{C}$ and let $e \in A$ be idempotent so that eAe is a subalgebra of A. Then it is possible to connect representations of A and eAe. Let $\mathcal{M}(A)$ and $\mathcal{M}(eAe)$ be the categories of A-modules and eAe-modules. There exist restriction and induction functors

$$r : \mathcal{M}(A) \to \mathcal{M}(eAe) \qquad \text{and} \qquad i : \mathcal{M}(eAe) \to \mathcal{M}(A)$$
$$Y \mapsto eY \qquad\qquad\qquad\qquad Z \mapsto A \underset{eAe}{\otimes} Z.$$

These have the following properties:

1. $r(i(Z)) = Z$,
2. r is exact.

Since r may annihilate some A-modules, it is not possible to recapture all representations of A from representations of eAe. However, if we let

$$\mathcal{M}(A, e) = \{Y \in \mathcal{M}(A) : Y = AeY\}$$

be the category of A-modules generated by e-fixed vectors, and denote irreducible A-modules by $\hat{A}$, then we have the following property.

3. $r : \hat{A} \cap \mathcal{M}(A, e) \to \widehat{eAe}$ is bijective.

Still there is the problem that $\mathcal{M}(A, e)$ may not be closed under taking submodules. However, the following are equivalent:

4. $i(\mathcal{M}(eAe)) = \mathcal{M}(A, e)$,
5. $\mathcal{M}(A, e)$ is closed under taking submodules,
6. $\mathcal{M}(eAe) \simeq \mathcal{M}(A, e)$.

Thus, according to statement (iii) above, it is possible to understand at least those irreducible representations of G that have K-fixed vectors by studying $\mathcal{H}(G//K)$.

Proposition 1.1. $\mathcal{H}(G//K)$ *has a basis consisting of* $f_g = \chi_{KgK}$ *for* $g \in K\backslash G/K$.

Proof. By definition, $\mathcal{H}(G//K)$ is generated by such functions. If

$$a_1 \chi_{Kg_1K} + \cdots + a_\ell \chi_{Kg_\ell K} = 0,$$

then evaluating at a representative of each double coset shows that $a_1 = \cdots = a_\ell = 0$ so such functions are also independent.

Suppose that G is unimodular. Notice that if we write

$$KgK = \cup_{i=1}^{m} k_i g K \text{ for } k_i \in K/(g^{-1}Kg \cap K)$$
$$= \cup_{j=1}^{n} Kg k_j \text{ for } k_j \in K/(gKg^{-1} \cap K),$$

then in fact $m = n = \mu(KgK)$ since right and left Haar measure are equal.

Proposition 1.2. *If* $f_x * f_y = \sum_z a_{xy}^z f_z$ *then the coefficients* $a_{xy}^z \in \mathbb{Z}$ *and thus*

$$\mu(KxK)\mu(KyK) = \sum_z a_{xy}^z \mu(KzK).$$

Proof. Writing $KxK = \cup_i k_i x K$ yields

$$f_x = \chi_{KxK} = \sum_i \chi_{k_i x K} = \sum_i \delta_{k_i x} * \chi_K.$$

If f is left K-invariant, then

$$f_x * f = \left(\sum_i \delta_{k_i x} * \chi_K \right) * f = \sum_i \delta_{k_i x} * (\chi_K * f) = \sum_i \delta_{k_i x} * f.$$

Since f_y is left K-invariant and can be written $f_y = \sum_j \delta_{\bar{k}_j y} * \chi_K$, this yields

$$f_x * f_y = \sum_{i,j} \delta_{k_i x} * \delta_{\bar{k}_j y} * \chi_K = \sum_z a_{xy}^z f_z.$$

The first sum can be written as an integral combination of characteristic functions of right K cosets. Partitioning the right cosets into double cosets yields a sum of characteristic functions of double cosets with integral coefficients. The second statement follows directly from the first.

Corollary 1.1. *If* $\mu(KxK)\mu(KyK) = \mu(KxyK)$ *then* $f_x * f_y = f_{xy}$.

2 Structure of p-adic $\mathrm{GL}(V)$ and $\mathrm{Sp}(V)$

After some preliminaries, we indicate how differently double cosets can behave in a non-unimodular group. See Example 2.1.

2.1 Preliminaries

Let k be a p-adic field, with ring of integers $\mathcal{O}$, and maximal ideal $\mathcal{P}$. Let π be a generator of $\mathcal{P}$. Let $\bar{k} = \mathcal{O}/\mathcal{P}$ be the *residue class field* of k. Let $q = \#(\bar{k})$ be the number of elements of $\bar{k}$. Let p be the *residual characteristic* of k, that is, the characteristic of $\bar{k}$. Then $q = p^a$, where a is the dimension of $\bar{k}$ over the prime field $\mathbb{F}_p$ of p elements.

Let $\bar{k}^\times$ denote the multiplicative group of $\bar{k}$. It is known that, for each element $\bar{x}$ of $\bar{k}^\times$, there is a unique $(q-1)$-th root of unity x in $\mathcal{O} \subset k$, such that the image of x in $\bar{k}$ is $\bar{x}$. Hence, by abuse of notation, we will henceforth understand by $\bar{k}^\times$ the group of $(q-1)$-th roots of unity of k, or the multiplicative group of $\bar{k}$, whichever makes sense in context.

Let $\mathcal{O}^\times$ be the group of units in $\mathcal{O}$. Note that $\mathcal{O}^\times = \mathcal{O} \setminus \mathcal{P}$ (set-theoretic difference). It is well-known that we can write

$$k^\times = \langle \pi \rangle \mathcal{O}^\times \quad \text{and} \quad \mathcal{O}^\times = \bar{k}^\times (1 + \mathcal{P}).$$

Here $\langle \pi \rangle$ is the cyclic group generated by π, $\bar{k}^\times$ is as above, and $1 + \mathcal{P} = \{1 + x : x \in \mathcal{P}\}$.

Example 2.1. Double cosets in a non-unimodular group.
Let

$$G = \left\{ \begin{bmatrix} a & b \\ 0 & 1 \end{bmatrix} : a \in k^\times, b \in k \right\},$$

$$K = \left\{ \begin{bmatrix} a & b \\ 0 & 1 \end{bmatrix} : a \in \mathcal{O}^\times, b \in \mathcal{O} \right\}, \text{ and}$$

$$z = \begin{bmatrix} \pi & 0 \\ 0 & 1 \end{bmatrix}.$$

In this case, a single left coset can equal arbitrarily many right cosets and vice versa. Since $z \begin{bmatrix} a & b \\ 0 & 1 \end{bmatrix} z^{-1} = \begin{bmatrix} a & \pi b \\ 0 & 1 \end{bmatrix}$, we know $zK \subseteq Kz$. Thus

$$\text{for } l \geq 0, \quad Kz^l K = Kz^l = \bigcup \begin{bmatrix} 1 & x \\ 0 & 1 \end{bmatrix} z^l K, \text{ and}$$

$$\text{for } l \leq 0, \quad Kz^l K = z^l K = \bigcup Kz^l \begin{bmatrix} 1 & x \\ 0 & 1 \end{bmatrix},$$

where in both cases $x \in \mathcal{O}/\pi^l \mathcal{O}$.

Let V be a vector space of dimension n over k. By a *lattice* in V, we mean a compact, open $\mathcal{O}$-module. Let Λ be a lattice in V. Then $\pi\Lambda$ is also a lattice and $\bar{\Lambda} = \Lambda/\pi\Lambda$ is a vector space over $\bar{k}$. Choose a basis $\bar{E} = \{\bar{e}_1, \ldots, \bar{e}_m\}$ for $\bar{\Lambda}$. Lift each element $\bar{e}_j$ of $\bar{\Lambda}$ to an element e_j of Λ.

Let x be any element of Λ. If $\bar{x}$ is the image of x in $\bar{\Lambda}$, then we may write

$$\bar{x} = \sum_{j=1}^{m} \bar{\beta}_j \bar{e}_j, \tag{1}$$

where the $\bar{\beta}_j$ are coefficients in $\bar{k}$. If we lift the coefficients $\bar{\beta}_j$ to elements of $\mathcal{O}$, then equation (1) says that $x - \sum_{j=1}^{m} \beta_j e_j = x_1$ belongs to $\pi\Lambda$. Replacing x with $\pi^{-1}x_1$ in equation (1), and letting $(\beta_1)_j$ be the corresponding coefficients, we see that $x - \sum_{j=1}^{m}(\beta_j + \pi(\beta_1)_j)e_j$ belongs to $\pi^2\Lambda$. Continuing in this fashion, we see that the set $E = \{e_j\}$ spans Λ as an $\mathcal{O}$-module, and in particular, spans V. It is easy to see that E must also be a linearly independent set, so it is a basis for V. Thus we may make the following statement.

Proposition 2.1. *Let $\Lambda \subset V$ be a lattice in the vector space V over k. Let $E = \{e_j\} \subset \Lambda$, and let $\bar{E} = \{\bar{e}_j\} \subset \bar{\Lambda} = \Lambda/\pi\Lambda$ be the set of images of the e_j in $\bar{\Lambda}$. If $\bar{E}$ is a basis for $\bar{\Lambda}$, then E is an $\mathcal{O}$-basis for Λ and a k-basis for V.*

2.2 Bruhat decomposition of GL(V)

We begin by stating some definitions that are needed for the two formulations of the Bruhat decomposition of $G = \mathrm{GL}(V)$ discussed below. Let V be a vector space of dimension n over k.

Definition 2.1. *A set $\mathcal{F} = \{U_1 \subset U_2 \subset \cdots \subset U_k\}$, of nested subspaces U_i of V, is called a flag in V. If $\dim U_j = j$ for $1 \le j \le k = n$, then F is called complete.*

Definition 2.2. *A* line decomposition *of V is a collection of lines ($=$ one-dimensional subspaces) L_j of V, such that V is a direct sum: $V = \oplus_j L_j$. Given a set, $\mathcal{C} = \{U_\alpha\}$, of subspaces of V, and a line decomposition $V = \oplus L_j$, the line decomposition is* compatible with $\mathcal{C}$ *if $U_\alpha = \oplus(L_j \cap U_\alpha)$ for all α.*

A parabolic subgroup of $\mathrm{GL}(V)$ can be viewed as the stabilizer of all subspaces in a flag $\mathcal{F}$, written as $P_{\mathcal{F}}$. In particular, a Borel subgroup $B = P_{\mathcal{F}_o}$ is the stabilizer of a complete flag. Even more particularly, if we choose a basis $E = \{e_1, \ldots, e_n\}$ and denote by V_k the space spanned by the first k elements of E, then the stabilizer of the complete flag $\mathcal{F}_o = \{V_1 \subset V_2 \subset \cdots \subset V_{n-1} \subset V_n\}$ is the subgroup of upper triangular matrices with respect to the basis E.

The notion of a line decomposition compatible with a flag $\mathcal{F}$ leads easily to an analogous notion of a compatible basis for V. The previous statement says that if a compatible basis is chosen and the matrices in the stabilizer of a complete flag are written in terms of this basis, then they are upper triangular. If $\mathcal{F}_o$ is a complete flag with compatible basis F, set $W = \mathrm{Stab}_{\mathrm{GL}(V)} F$. This is the Weyl group, which for $\mathrm{GL}(V)$ is the symmetric group on n elements.

Theorem 2.1 (Bruhat Decomposition). *For any Borel subgroup, B,* $GL(V) = BWB$.

This theorem has the following refinements. Let A be the subgroup of $GL(V)$ for which the basis E is an eigenbasis, i.e., the stabilizer of all individual lines in the line decomposition corresponding to E. Let U be the unipotent radical of B (the set of elements of B which act trivially on the quotients U_{j+1}/U_j).

Theorem 2.2. $GL(V) = UWAU = UAWU = U^w AWU^{w'}$, *for any* $w, w' \in W$, *where* $U^w = wUw^{-1}$.

This could also be written as $G = U\widetilde{W}U$ where the affine Weyl group, $\widetilde{W} = WA$ is the semidirect product of W and A, or the stabilizer of the overall line decomposition of V.

The Bruhat Decomposition also has the following equivalent more geometric reformulation.

Theorem 2.3. $G = BWB$ *if and only if for any two flags, $\mathcal{F}_1$ and $\mathcal{F}_2$, there exists a line decomposition of V compatible with both $\mathcal{F}_1$ and $\mathcal{F}_2$.*

Proof. First suppose that the geometric version holds. Fix an element $g \in GL(V)$ and a flag $\mathcal{F}_1$ with $B = \mathrm{Stab}_{GL(V)}\mathcal{F}_1$. Choose a compatible basis F_1 for $\mathcal{F}_1$ and let $W = \mathrm{Stab}_{GL(V)}F_1$. Set $g(\mathcal{F}_1) = \mathcal{F}_2$ where $\mathcal{F}_2$ has compatible basis $F_2 = g(F_1)$. Choose a basis E compatible with $\mathcal{F}_1$ and $\mathcal{F}_2$ and choose $b_1 \in B$ such that $b_1(F_1) = E$. Since E is compatible with $\mathcal{F}_1$ and $\mathcal{F}_2$ and since $b_1^{-1}(E) = F_1$, F_1 is a basis compatible with both $b_1^{-1}(\mathcal{F}_1) = \mathcal{F}_1$ and $b_1^{-1}(\mathcal{F}_2) = b_1^{-1}g(\mathcal{F}_1) = \mathcal{F}_3$. This means we can find $w \in W$ so that $w(\mathcal{F}_1) = \mathcal{F}_3$, i.e., so that the kth subspace in the flag $\mathcal{F}_3$ is the span of the first k elements in the reordered basis $w(F_1)$. Then $w^{-1}b_1^{-1}g = b_2 \in \mathrm{Stab}_{GL(V)}\mathcal{F}_1 = B$, so $g \in BWB$. The idea is illustrated by the following commutative diagram.

$$
\begin{array}{ccc}
\mathcal{F}_1 & \xrightarrow{\ g\ } & \mathcal{F}_2 \\
{\scriptstyle b_2}\downarrow & & \downarrow{\scriptstyle b_1^{-1}} \\
\mathcal{F}_1 & \xrightarrow{\ w\ } & \mathcal{F}_3
\end{array}
$$

Now suppose we are given any two flags and expand them if necessary to complete flags, $\mathcal{F}_1$ and $\mathcal{F}_2$. As before, let $B = \mathrm{Stab}_{GL(V)}\mathcal{F}_1$ and let $W = \mathrm{Stab}_{GL(V)}F_1$ where F_1 is a compatible basis for $\mathcal{F}_1$. Find $g \in GL(V)$ such that $g(\mathcal{F}_1) = \mathcal{F}_2$ and write $g = b_1 w b_2$. We claim $b_1(F_1) = E$ is a basis compatible with both $\mathcal{F}_1$ and $\mathcal{F}_2$. Since $b_1 \in \mathrm{Stab}_{GL(V)}\mathcal{F}_1$, E is compatible with $\mathcal{F}_1$. Now consider $b_1 w b_1^{-1}(E)$. On the one hand, this equals $b_1 w(F_1)$, which is some reordering of the vectors in E. But then that equals $g b_2^{-1}(F_1)$, and since $b_2^{-1} \in \mathrm{Stab}_{GL(V)}\mathcal{F}_1$, $b_2^{-1}(\mathcal{F}_1)$ is compatible with $\mathcal{F}_1$ and gets sent by g to a basis compatible with $\mathcal{F}_2$.

Thus, to establish the Bruhat Decomposition, we will prove the following result.

Theorem 2.4 (Bruhat Decomposition, geometric version). *Given any two flags, $\mathcal{F}_1$ and $\mathcal{F}_2$, there exists a line decomposition of V compatible with both $\mathcal{F}_1$ and $\mathcal{F}_2$.*

Proof. Proceed by induction on the dimension of V, with the case of dimension one being trivial. Given any two flags, expand them if necessary to complete flags, $\mathcal{F}_1 = \{U_1 \subset U_2 \subset \cdots \subset U_n\}$ and $\mathcal{F}_2 = \{W_1 \subset W_2 \subset \cdots \subset W_n\}$, and let k_1 be the smallest index such that U_1 is contained in W_{k_1}. Let the first line in the line decomposition be $L_1 = U_1$. Let Z be any complement of W_{k_1} in V and set $V_1 = W_{k_1-1} + Z$.

V_1 is of dimension $n-1$ and since $U_1 + V_1 = V$, then $U_j + V_1 = V$ as well, for any $j \geq 1$. This forces $\dim U_j \cap V_1 = \dim U_j + \dim V_1 - \dim(U_j + V_1) = j-1$. Thus for $2 \leq j \leq n$ the $U_j \cap V_1$ form a complete flag in V_1.

Next, note that

$$W_j \cap V_1 = \begin{cases} W_j & 1 \leq j \leq k_1 - 1 \\ W_{k_1-1} + (W_j \cap Z) & k_1 \leq j \leq n \end{cases}$$

and therefore

$$\dim(W_j \cap V_1) = \begin{cases} j & 1 \leq j \leq k_1 - 1 \\ j-1 & k_1 \leq j \leq n \end{cases}$$

where the result is the same for both $j = k_1 - 1$ and $j = k$. Thus the $W_j \cap V_1$ yield a complete flag in V_1 as well. By induction, there exists a line decomposition $V_1 = \bigoplus_{j=2}^{n} L_j$ compatible with these two complete flags. Using dimension arguments, it is possible to show that $\bigoplus_{j=1}^{n} L_j$ is a line decomposition of V compatible with the two original flags. If we set $\sigma(1) = k_1$, then induction yields a permutation σ corresponding to the line decomposition. (For example, $\sigma(2)$ would be the smallest index such that $U_2 \cap V_1$ is contained in $W_{\sigma(2)} \cap V_1$.) This shows how to reorder the lines so that $W_k = \bigoplus_{j=1}^{k} L_{\sigma^{-1}(j)}$.

2.3 Iwahori-Bruhat decomposition of GL(V)

Lattice Flags For p-adic groups, there is a lattice analog of flags and an analog involving lattices of the Bruhat decomposition.

Definition 2.3. *A set $\mathcal{L}$ of lattices is called a lattice flag if and only if*
(a) it is nested, i.e., totally ordered by inclusion, and
(b) it is invariant under multiplication by $k^{\times}$.

Let $\mathcal{L}$ be a lattice flag, and let Λ_o be a lattice in $\mathcal{L}$. Since $\mathcal{L}$ is invariant under multiplication by scalars, if $x = \pi^a u$ is an element of k, where $u \in \mathcal{O}^\times$ is a unit, then $x\Lambda_o = \pi^a \Lambda_o$, since $u\Lambda_o = \Lambda_o$. Thus, to satisfy (a), it is enough to know, for any lattice $\Lambda_o \in \mathcal{L}$, that $\pi^{\pm 1}\Lambda_o$ is also contained in $\mathcal{L}$.

Let Λ' be any other element of $\mathcal{L}$. For sufficiently large positive exponents a, we will have $\pi^a \Lambda' \subset \Lambda_o$. Choose such an a to be as small as possible. By minimality we then have $\pi^a \Lambda' \not\subset \pi\Lambda_o$. Since $\mathcal{L}$ is totally ordered by inclusion, this implies that $\pi^a \Lambda' \supset \pi\Lambda_o$. Reduction modulo π attaches to $\pi^a \Lambda'$ a subspace $U_{\Lambda'} \subset \overline{\Lambda}_o = \Lambda_o/\pi\Lambda_o$. It is clear that $\pi^a \Lambda'$ can be recovered from $U_{\Lambda'}$, as the unique lattice containing $\pi\Lambda_o$, and reducing modulo $\pi\Lambda_o$ to $U_{\Lambda'}$. From $\pi^a \Lambda'$, all multiples of Λ' can be recovered.

Let Λ'' be a third lattice in $\mathcal{L}$. Repeating the reasoning of the previous paragraph, we find an exponent b such $\pi\Lambda_o \subset \pi^b \Lambda'' \subset \Lambda_o$, and a subspace $U_{\Lambda''} \subset \overline{\Lambda}_o$ such that $\pi^b \Lambda''/\pi\Lambda_o = U_{\Lambda''}$. Appealing to the ordering with respect to inclusion of $\mathcal{L}$, we see that the two subspaces $U_{\Lambda'}$ and $U_{\Lambda''}$ must also be related by inclusion, that is, one must contain the other.

If we extend this analysis to all lattices of $\mathcal{L}$, we see that Λ gives rise to a collection $\{U_{\tilde{\Lambda}} : \tilde{\Lambda} \in \mathcal{L}\}$ of subspaces of $\overline{\Lambda}_o$. Since an element of $\mathcal{L}$ can be recovered up to multiples from its associated subspace of $\overline{\Lambda}_o$, the collection $\{U_{\tilde{\Lambda}} : \tilde{\Lambda} \in \mathcal{L}\}$ determines $\mathcal{L}$. The ordering with respect to inclusion of $\mathcal{L}$ translates to the same relation between the $U_{\tilde{\Lambda}}$. Hence the $U_{\tilde{\Lambda}}$ form a flag in $\overline{\Lambda}_o$. Thus, $\mathcal{L}$ determines and is determined by a flag in $\overline{\Lambda}_o$.

Conversely, given a lattice Λ_o and a flag $\{U_\alpha\}$ in $\overline{\Lambda}_o = \Lambda_o/\pi\Lambda_o$, we can form the lattices Λ_α such that $\Lambda_\alpha \supset \pi\Lambda_o$, and $\Lambda_\alpha/\pi\Lambda_o = U_\alpha$. Then taking all multiples $\pi^m \Lambda_\alpha$ of these lattices, it is not hard to see that we obtain a lattice flag. Thus, we conclude that the collection of all lattice flags containing a given lattice Λ_o is in bijection with the collection of all flags in the residue class vector space $\overline{\Lambda}_o$.

There is an obvious notion of maximal lattice flag. It is clear that in the above correspondence between lattice flags containing Λ_o and flags in $\overline{\Lambda}_o$, maximal lattice flags correspond to maximal flags in $\overline{\Lambda}_o$. One consequence is that every lattice flag can be extended to a maximal lattice flag. Also, a lattice flag is maximal if and only if the quotients, Λ'/Λ'', of consecutive elements $\Lambda'' \subset \Lambda'$ of the flag are one-dimensional over the residue class field $\overline{k}$.

Stabilizers of Lattices Let $\mathrm{GL}(V)$ be the group of linear automorphisms of the vector space V. For a lattice $\Lambda \subset V$, let $K(\Lambda)$ ($= K$ when Λ is understood) be the subgroup of $\mathrm{GL}(V)$ consisting of automorphisms of Λ. That is $K(\Lambda) = \{g \in \mathrm{GL}(V) : g(\Lambda) = \Lambda\}$.

Proposition 2.2. *There is a unique conjugacy class of maximal compact subgroups of* $\mathrm{GL}(V)$, *consisting of the stabilizers* $K(\Lambda)$ *of lattices* Λ. *In other words, every maximal compact subgroup of* $\mathrm{GL}(V)$ *stabilizes a lattice, and all stabilizers of lattices are conjugate to each other in* $\mathrm{GL}(V)$.

Proof. For any fixed lattice Λ, the stabilizer $K(\Lambda)$ is a compact and open subgroup of $\mathrm{GL}(V)$. If Λ' is another lattice, then we can find $g \in \mathrm{GL}(V)$ such that $g(\Lambda) = \Lambda'$. This can be seen, for example, by choosing bases for Λ and for Λ', and letting g be the mapping which takes one basis to the other. Then $K(\Lambda') = g(K(\Lambda))g^{-1}$, so $K(\Lambda)$ and $K(\Lambda')$ are conjugate.

Let H be any compact subgroup of $\mathrm{GL}(V)$. Since $K(\Lambda)$ is open, the intersection $H \cap K(\Lambda)$ has finite index in H. This means that the lattices $\{h(\Lambda) : h \in H\}$ form a finite set. Hence the sum of such lattices will again be a lattice $\tilde{\Lambda}$ in V, and will evidently be stabilized by H. Hence, $H \subset K(\tilde{\Lambda})$.

Consider the structure of $K(\Lambda)$. Evidently, $K(\Lambda)$ stabilizes all the multiples $\pi^a \Lambda$ of Λ. In particular, any $h \in K(\Lambda)$ stabilizes $\pi\Lambda$, and therefore descends to define a mapping of the quotient $\overline{\Lambda} = \Lambda/\pi\Lambda$. This gives us a homomorphism from $K(\Lambda)$ to $\mathrm{GL}(\overline{\Lambda})$. By choosing a basis for Λ, we can see that this mapping is surjective, so that we have an exact sequence

$$1 \to K_1(\Lambda) \to K(\Lambda) \to \mathrm{GL}(\overline{\Lambda}) \to 1.$$

An element y in $K(\Lambda) = K$ is in K_1 if and only if $\overline{y}(\overline{x}) = \overline{x}$ for all x in Λ. This means that $\overline{y}(\overline{x}) - \overline{x}$ belongs to $\pi\Lambda$. In other words, $(1 - y)(\Lambda) \subset \pi\Lambda$, or $z = \pi^{-1}(y - 1)$ preserves Λ. Let $\mathrm{End}(\Lambda)$ denote the ring of linear maps of V which preserve Λ. It is a compact open subring of $\mathrm{End}(V)$ - a lattice in $\mathrm{End}(V)$. We have shown that

$$K_1(\Lambda) = 1 + \pi\mathrm{End}(\Lambda).$$

We define the *level m principal congruence subgroup* of $K(\Lambda)$ by

$$K_m(\Lambda) = 1 + \pi^m \mathrm{End}(\Lambda).$$

This is a normal, open subgroup of $K(\Lambda)$. As $m \to \infty$, the groups K_m run through a neighborhood basis of the identity in K. Also, the map $y \to 1 + y$ defines the first of the following isomorphisms:

$$K_m(\Lambda)/K_{m+1}(\Lambda) \simeq \pi^m \mathrm{End}(\Lambda)/\pi^{m+1}\mathrm{End}(\Lambda)$$
$$\simeq \mathrm{End}(\Lambda)/\pi\mathrm{End}(\Lambda) \simeq \mathrm{End}(\overline{\Lambda}).$$

From this, we see that the quotient groups K_m/K_{m+1} are vector spaces over $\overline{k}$, and in particular are abelian p-groups.

As a by product of the description given above, we see that $K(\Lambda)$ acts transitively on $\Lambda - \pi\Lambda$. Therefore, it will act transitively on $\pi^c\Lambda - \pi^{c+1}\Lambda$ for any integer c. This implies the following fact:

Lemma 2.1. *A lattice stabilized by $K(\Lambda)$ has the form $\pi^c\Lambda$ for some integer c.*

Recall now some generalities about pro-finite groups. Let H be a *pro-finite group*, i.e., a compact, totally disconnected group. Since H is totally disconnected, every point has a neighborhood basis of open and closed sets. Thus, H has a neighborhood basis of the identity consisting of open and closed subsets X_α. For a given X_α, the stabilizer H_α of X_α under left multiplication will be open and closed, and contained in X_α itself. Since it is open, it is of finite index in H, whence it has only finitely many conjugates in H. Thus, the intersection of all these conjugate subgroups will again be open, and will be an open, normal subgroup of H, and again clearly contained in X_α. Thus, H allows a neighborhood basis of the identity element, consisting of open, normal subgroups. Call them N_α. This exhibits H as the inverse or projective limit of its finite quotients H/N_α:

$$H = \varprojlim H/N_\alpha.$$

Hence the term pro-finite. We can think of the order of H as being a product $2^{m_2} 3^{m_3} 5^{m_5} \cdots$, where the exponents are either whole numbers or $+\infty$. As N_α shrinks, an exponent in the order of H/N_α gets larger and larger, and either remains bounded, stabilizing at m_p, or increases without bound, in which case $m_p = \infty$.

There is a Sylow theory for pro-finite groups. A *Sylow subgroup* H_p of H is a subgroup which projects to a Sylow subgroup in any finite quotient H/N_α.

Proposition 2.3. *1. Sylow subgroups H_p of a pro-finite group H exist.*
2. All Sylow p-subgroups of H are conjugate.
3. The order of H_p is p^{m_p}.

We refer to a group in which only one exponent m_p is positive as a *pro-p group*. Evidently, the Sylow subgroup H_p of H is a maximal pro-p subgroup of H.

Return now to the stabilizer $K(\Lambda)$ of a lattice. From our description above of the congruence subgroup $K_1(\Lambda)$, we can see that it is a pro-p subgroup. It follows that all the exponents of $K(\Lambda)$ are finite except for m_p. Also the Sylow p-subgroups of $K(\Lambda)$ will be the inverse images in $K(\Lambda)$ of the Sylow p-subgroups of $\mathrm{GL}(\overline{\Lambda})$. It is well known that these are the groups of "unipotent upper triangular matrices". Precisely, if $\overline{\mathcal{F}}$ is a maximal flag in $\overline{\Lambda}$, then the subgroup of $\mathrm{GL}(\overline{\Lambda})$ which stabilizes $\overline{\mathcal{F}}$ and acts trivially on the quotients of successive elements of $\overline{\mathcal{F}}$ is a Sylow p-subgroup. We will denote by $J = J(\Lambda, \overline{\mathcal{F}})$ the stabilizer in $K(\Lambda)$ of $\overline{\mathcal{F}}$, and by $J_1 = J_1(\Lambda, \overline{\mathcal{F}})$ the Sylow p-subgroup of J. Then in fact, J is the normalizer in K of J_1, and J/J_1 is abelian and isomorphic to $(\overline{k}^\times)^n$.

As we have discussed above, there is naturally associated to the flag $\overline{\mathcal{F}}$ a maximal lattice flag $\mathcal{L}$. The stabilizer $J(\Lambda, \overline{\mathcal{F}})$ is equal to the stabilizer $J(\mathcal{L})$. The subgroup $J_1(\Lambda, \overline{\mathcal{F}}) = J_1(\mathcal{L})$ consists of elements which act trivially on the quotients of consecutive elements of $\mathcal{L}$. We call $J(\mathcal{L})$ an *Iwahori subgroup* of $\mathrm{GL}(V)$. More generally, if $\mathcal{M}$ is any lattice flag in V, then the stabilizer of (all the lattices of) $\mathcal{M}$ is called a *parahoric subgroup* of $\mathrm{GL}(V)$. From our discussion of the relation between lattice flags containing Λ and flags in $\overline{\Lambda}$, we know that if Λ belongs to $\mathcal{M}$, then $K_1(\Lambda) \subset J(\mathcal{M})$, and the quotient $J(\mathcal{M})/K_1(\Lambda)$ is identified to the stabilizer of a flag in $\overline{\Lambda}$ - a parabolic subgroup of $\mathrm{GL}(\overline{\Lambda})$.

Combining the discussions of $K(\Lambda)$ and of Sylow subgroups, we may make the following conclusion.

Corollary 2.1. *There is a unique conjugacy class of maximal compact pro-p subgroups of $\mathrm{GL}(V)$, consisting of the $J_1(\mathcal{L})$, for all maximal lattice flags $\mathcal{L}$.*

Iwahori-Bruhat Decomposition We now turn to the analog for lattice flags of the Bruhat decomposition.

Definition 2.4. *Let $\mathcal{L} = \{\Lambda_\alpha\}$ be a collection of lattices, and let $V = \oplus_j L_j$ be a line decomposition of V. We say that the line decomposition is* compatible *with $\mathcal{L}$, if for each lattice $\Lambda_\alpha \in \mathcal{L}$, we have $\Lambda_\alpha = \oplus_j(\Lambda_\alpha \cap L_j)$.*

Proposition 2.4. Iwahori-Bruhat decomposition, geometric ver–sion *Let $\mathcal{L}$ and $\mathcal{M}$ be two lattice flags. Then there is a line decomposition $V = \oplus_j L_j$ compatible with both $\mathcal{L}$ and $\mathcal{M}$.*

Proof. Select a lattice Λ_o of $\mathcal{L}$. As we have seen, $\mathcal{L}$ is then associated to and determined by a flag $\overline{\mathcal{F}}(\mathcal{L})$ in $\overline{\Lambda}_o$. We will construct another flag in $\overline{\Lambda}_o$ using $\mathcal{M}$. For each lattice M of $\mathcal{M}$, set $\tilde{M} = (M \cap \Lambda_o) + \pi\Lambda_o$. This is a lattice between $\pi\Lambda_o$ and Λ_o, so that it defines and is determined by a subspace $U(M)$ of $\overline{\Lambda}_o$. For sufficiently small elements of $\mathcal{M}$, the subspace $U(M)$ will be zero, and for sufficiently large elements of $\mathcal{M}$, it will be all of $\overline{\Lambda}_o$. Since the elements of $\mathcal{M}$ are totally ordered by inclusion the $U(M)$ will define a flag in $\overline{\Lambda}_o$. Call this flag $\overline{\mathcal{G}}(\mathcal{M})$. Since the quotient of successive elements of $\mathcal{M}$ is just a line over $\overline{k}$, the flag $\overline{\mathcal{G}}(\mathcal{M})$ will be a maximal flag.

According to the Bruhat decomposition for $\mathrm{GL}(\overline{\Lambda})$, we can find a line decomposition of $\overline{\Lambda}_o$ which is compatible with both flags $\overline{\mathcal{F}}(\mathcal{L})$, and $\overline{\mathcal{G}}(\mathcal{M})$. Let $\{\overline{z}_j\}$ be a basis for $\overline{\Lambda}_o$ selected from the lines of the line decomposition. Then, since the flag $\overline{\mathcal{F}}$ is defined by lattices between Λ_o and $\pi\Lambda_o$, we may choose *any* elements $\{z_j\}$ lifting the $\overline{z}_j$ to Λ_o, and these elements will be basis elements for lines, such that the resulting line decomposition of V will be compatible with $\mathcal{L}$. On the other hand, each $\overline{z}_j$ also spans the quotient of two spaces $U(M_1)$ and $U(M_2)$, where $M_1 \subset M_2$ are two successive elements of $\mathcal{M}$. Thus, M_1 is the largest element of $\mathcal{M}$ such that $U(M_1)$ does not contain $\overline{z}_j$, and M_2 is the smallest element of $\mathcal{M}$ such that $U(M_2)$ does contain it. Thus, we may lift $\overline{z}_j$ to an element z_j belonging to M_2. We claim that if this is done, the line decomposition defined by the z_j is compatible also with $\mathcal{M}$.

Indeed, consider any lattice M in $\mathcal{M}$. The multiples $\pi^c M$ of M define a subflag of $\overline{\mathcal{G}}(\mathcal{M})$. According to our definitions, we have

$$U(\pi^c M)/U(\pi^{c+1} M) \simeq (\pi^c M \cap \Lambda_o)/(\pi^c M \cap \pi \Lambda_o) + (\pi^{c+1} M \cap \Lambda_o).$$

An appropriate subset of the z_j belong to $\pi^c M \cap \Lambda_o$, and reduce modulo $\pi \Lambda_o$ to define a basis of the quotient.

We observe that the formation of the quotients above is essentially symmetric in M and Λ_o. In particular, we have

$$(\pi^c M \cap \Lambda_o)/((\pi^c M \cap \pi \Lambda_o) + (\pi^{c+1} M \cap \Lambda_o))$$
$$= \pi^c[(M \cap \pi^{-c} \Lambda_o)/((M \cap \pi^{-c+1} \Lambda_o) + (\pi M \cap \pi^{-c} \Lambda_o))]$$
$$\simeq \quad (M \cap \pi^{-c} \Lambda_o)/((M \cap \pi^{-c+1} \Lambda_o) + (\pi M \cap \pi^{-c} \Lambda_o)).$$

Thus, for each z_j, there is an appropriate exponent c_j such that the elements $\pi^{-c_j} z_j$, will be a basis for M, and when reduced modulo πM will provide a basis compatible with the flag in $\overline{M}$ defined by the multiples of Λ_o. Since this is true for any lattice in $\mathcal{M}$, this means that the elements z_j define a line decomposition compatible with $\mathcal{M}$. Thus the proposition is proved.

Remark 2.1. 1. As in the case of the ordinary Bruhat Decomposition, this geometric version of the Iwahori-Bruhat Decomposition has a group theoretic formulation. As above, let $J(\mathcal{L}) = J$ be the stabilizer of the maximal lattice flag $\mathcal{L}$. Let Λ_o be a lattice in $\mathcal{L}$, and let $K(\Lambda_o)$ be the stabilizer of Λ_o. Let $V = \oplus_j L_j$ be a line decomposition of V compatible with $\mathcal{L}$. Let A be the group of transformations which stabilize all the lines L_j, and let $\tilde{W} = AW$ be the affine Weyl group of transformations which stabilize the collection $\{L_j\}$ (but which may permute the L_j among themselves). Then the group-theoretic version of the Iwahori-Bruhat Decomposition is

$$\mathrm{GL}(V) = J(\mathcal{L})\tilde{W} J(\mathcal{L}).$$

2. Here we are abusing language by calling AW the "affine Weyl group". Correctly speaking, the affine Weyl group is AW/A_0, where $A_0 = A \cap J(\mathcal{L})$ is the maximal compact subgroup of A.

3. There is also a related description of the double cosets of the maximal compact subgroup $K(\Lambda_o)$:

$$\mathrm{GL}(V) = K(\Lambda_o)AK(\Lambda_o).$$

This is sometimes called the *Cartan decomposition.* It follows easily from the Iwahori-Bruhat decomposition. It also exists in a slightly finer version, in which A is replaced by a subsemigroup A^{0+} (see Section 3), which forms a fundamental domain for the action of W on A.

2.4 Bruhat decomposition of $\mathrm{Sp}(V)$

Let V be a vector space of dimension $2n$ equipped with a symplectic form $\langle\,,\,\rangle$. This allows us to define a map, denoted by $^\sharp : \mathrm{End}(V) \to \mathrm{End}(V)$, such that $\langle Tu, v \rangle = \langle u, T^\sharp v \rangle$. This map has the following properties:

1. $\sharp$ is linear,
2. $(T^\sharp)^\sharp = T$,
3. $(ST)^\sharp = T^\sharp S^\sharp$.

The symplectic group can be defined as

$$\mathrm{Sp}(V) = \{g \in \mathrm{GL}(V) \mid \langle gu, gv \rangle = \langle u, v \rangle = \langle u, g^\sharp g v \rangle \text{ or } g^\sharp g = I\},$$

with Lie algebra

$$\mathfrak{sp}(V) = \{T \in \mathrm{End}(V) \mid T^\sharp = -T\}.$$

There is the usual notion of duality with respect to the symplectic form. For a subspace U of V, define

$$U^\perp = \{v \in V \mid \langle v, u \rangle = 0 \text{ for all } u \in U\}.$$

Then $(U^\perp)^\perp = U$ and $\dim U + \dim U^\perp = \dim V$. We say U is *isotropic* if $\langle \, , \, \rangle|_U = 0$, which is equivalent to $U \subset U^\perp$. If all subspaces in a flag are isotropic then we say the flag is isotropic. If a flag has the form $\mathcal{F} = \{U_1 \subset U_2 \subset \cdots \subset U_\ell\}$, where $U_i^\perp = U_j$ then the flag is called *self-dual*. Notice that then in fact $j = \ell - i + 1$. There is a bijection between isotropic flags and self-dual flags, obtained by adding to the isotropic flag the orthogonal subspaces of the subspaces in the flag. We will refer to a self-dual flag with $\ell = 2n - 1$ (so $U_i^\perp = U_{2n-i}$) as *complete*. Such flags have the form $\mathcal{F} = \{U_1 \subset U_2 \subset \cdots \subset U_n \subset U_{n-1}^\perp \subset U_{n-2}^\perp \subset \cdots \subset U_1^\perp\}$, and are also called *symplectic* flags.

Let $E = \{e_j, f_j : 1 \leq j \leq n\}$ be a symplectic basis for V. This means that $\langle e_i, e_j \rangle = 0 = \langle f_i, f_j \rangle$, and $\langle e_i, f_j \rangle = \delta_{ij}$, where δ_{ij} is Kronecker's delta. Let B be the stabilizer of the complete symplectic flag associated to E, and let W denote the Weyl group, the stabilizer of the basis E modulo signs (meaning, that the set $\pm E$ is stable under W, and W acts by permutations on the set of pairs $\{\pm e_j\}.\{\pm f_j\}$.

Theorem 2.5 (Bruhat Decomposition). *For any Borel subgroup, B,* $\mathrm{Sp}(V) = BWB$.

Once again there are various refinements. Let U be the unipotent radical of B (the set of elements of B which act trivially on the quotients U_{j+1}/U_j). Let A be the subgroup that fixes every line ke_i or kf_i and let $\widetilde{W}$ denote the affine Weyl group, the stabilizer of the set of lines $\{ke_i, kf_i\}$. Then $\widetilde{W} = WA$ (but this is no longer a semidirect product as it was in the case of $\mathrm{GL}(V)$).

Theorem 2.6. $\mathrm{Sp}(V) = BWU = U\widetilde{W}U$.

There is again an equivalent geometric reformulation. Let $\mathcal{F}_1$ and $\mathcal{F}_2$ be two complete, self-dual flags in V. We know from the Bruhat theory for $\mathrm{GL}(V)$

that we can find a line decomposition of V compatible with both $\mathcal{F}_1$ and $\mathcal{F}_2$. However, this does us little good in understanding the structure of $\mathrm{Sp}(V)$, because $\mathrm{Sp}(V)$ does not act transitively on the set of line decompositions of V. In the context of $\mathrm{Sp}(V)$, we need to capitalize on the self-dual structure of the $\mathcal{F}_j$ to show that we can find a line decomposition compatible with the symplectic structure. Let $V = \oplus_j L_j$ be a line decomposition of V. We will call it *symplectic* if, for each index j, there is a unique index j' such that $L_{j'}$ is paired non-trivially with L_j. That is, L_j is orthogonal to all but one of the other L_k. Then the planes $L_j \oplus L_{j'}$ decompose V into an orthogonal direct sum of non-degenerate planes. Equivalently, we can find basis elements for the lines L_j such that they form a symplectic basis for V.

Theorem 2.7 (Bruhat Decomposition, geometric version). *Given any two self-dual flags, $\mathcal{F}_1$ and $\mathcal{F}_2$, there exists a symplectic line decomposition of V compatible with both $\mathcal{F}_1$ and $\mathcal{F}_2$.*

Proof. Fix a symplectic form on V. Proceed by induction on the dimension of V, with the case of dimension two being straightforward. Given two self-dual flags, expand them if necessary to obtain complete flags, $\mathcal{F}_1 = \{U_1 \subset U_2 \subset \cdots \subset U_{2n-1}\}$ and $\mathcal{F}_2 = \{W_1 \subset W_2 \subset \cdots \subset W_{2n-1}\}$, where $U_j^\perp = U_{2n-j}$ and $W_j^\perp = W_{2n-j}$. Define $k_1 = \sigma(1)$ as the smallest index such that U_1 is contained in W_{k_1}. Then taking complements, $W_{2n-k_1} \subset U_{2n-1}$ and $2n - k_1$ is the largest such index. Choose a line $L \subset W_{2n-k_1+1}$ complimentary to W_{2n-k_1}. Notice that this means $W_{k_1-1} \subset L^\perp$ and $k_1 - 1$ is the largest such index. Since U_1 and L are paired nontrivially under the symplectic form and since U_1 is isotropic, $P = U_1 \oplus L$ is a hyperbolic plane and $P^\perp$ is a $(2n-2)$-dimensional subspace of V.

We claim that the $U_j \cap P^\perp$ and the $W_j \cap P^\perp$ form symplectic flags in $P^\perp$. First we show they are self dual, i.e., that

$$(U_j \cap P^\perp)^\perp \cap P^\perp = U_{2n-j} \cap P^\perp \tag{2}$$

or equivalently, that

$$(U_{2n-j} + P) \cap P^\perp = U_{2n-j} \cap P^\perp, \tag{3}$$

and similarly for W_j. For simplicity of notation, set $2n - j = i$. For the U_i case, choose $x = u + \ell \in U_i + P$, with $u \in U_i$ and $\ell \in L$. (Note that $U_1 \subset U_i$, so that $U_i + P = U_i + L$.) Suppose $x \in P^\perp$. Then $u + \ell \perp U_1$, but $u \in U_i \subset U_{2n-1} = U_1^\perp$, so $\ell \perp U_1$. Since L and U_1 are paired nontrivially, this forces $\ell = 0$ and $x = u \in U_i$. Thus $(U_i + P) \cap P^\perp \subset U_i \cap P^\perp$ and $(U_i + P) \cap P^\perp = U_i \cap P^\perp$.

Similarly, take $x = w + u_1 + \ell \in W_i + P$. We want to show that if $x \in P^\perp$ then $x \in W_i$. There are four slightly different cases depending on the relative positions of i, k_1 and $2n - k_1 + 1$. Set

$$a = \min\{k_1 - 1, 2n - k_1\}$$
$$b = \max\{k_1, 2n - k_1 + 1\}.$$

Case 1, $i \leq a$: If $x \in P^{\perp}$, then $w + u_1 + \ell \perp U_1$ and since $w \in W_{2n-k_1} \subset U_1^{\perp}$ and $u_1 \in U_1 \subset U_1^{\perp}$, we see $\ell = 0$. But then $x = w + u_1$ is also in $L^{\perp}$ and $w \in W_{k_1-1} \subset L^{\perp}$ implies $u_1 = 0$ so $x \in W_i$.

Case 2, $k_1 \leq i \leq 2n - k_1$: Since $U_1 \subset W_i$, we can write $x = w + \ell$. If $x \in P^{\perp}$ then $w + \ell \perp U_1$, but $w \in W_{2n-k_1} \subset U_1^{\perp}$, so $\ell = 0$.

Case 3, $2n - k_1 + 1 \leq i \leq k_1 - 1$: Since $L \subset W_i$, we can write $x = w + u_1$. If $x \in P^{\perp}$, then $w + u_1 \perp L$, but $w \in W_{k_1-1} \subset L^{\perp}$, so $u_1 = 0$.

Case 4, $i \geq b$: Since $U_1 \subset W_i$ and $L \subset W_i$, $P^{\perp} \subset W_i$ so certainly $(W_i + P) \cap P^{\perp} \subset W_i \cap P^{\perp}$.

To see that both $U_i \cap P^{\perp}$ and $W_i \cap P^{\perp}$ form complete flags, notice that

$$U_i = (U_i \cap P) \oplus (U_i \cap P^{\perp})$$
$$W_i = (W_i \cap P) \oplus (W_i \cap P^{\perp}).$$

These decompositions are in fact equivalent to the equalities

$$(U_i + P) \cap P^{\perp} \subset U_i \cap P^{\perp}$$
$$(W_i + P) \cap P^{\perp} \subset W_i \cap P^{\perp},$$

as can be shown directly. The decompositions make it clear that for $1 \leq i \leq 2n - 1$, $\dim(U_i \cap P^{\perp}) = i - 1$, and

$$\dim(W_i \cap P^{\perp}) = \begin{cases} i & \text{if } i \leq a \\ i - 1 & \text{if } a < i \leq b \\ i - 2 & \text{if } b < i. \end{cases}$$

Now by induction, find a symplectic line decomposition $P^{\perp} = \bigoplus_{j=3}^{2n} L_j$ compatible with these two flags. Setting $L_1 = U_1$ and $L_2 = L$, it is possible to show, using dimension arguments, that $\bigoplus_{j=1}^{2n} L_j$ is a symplectic line decomposition of V compatible with the two original symplectic flags.

2.5 Iwahori-Bruhat decomposition of $\mathrm{Sp}(V)$

Self-dual Lattice Flags Let Λ be a lattice in the symplectic space V. The *dual lattice* Λ^* is defined by

$$\Lambda^* = \{y \in V : \langle y, x \rangle \in \mathcal{O}, \text{for every } x \in \Lambda\}.$$

We have defined a lattice flag in Section 2.3. We will call a lattice flag $\mathcal{L}$ *self-dual* if, for every lattice Λ in $\mathcal{L}$, the dual lattice Λ^* is also in $\mathcal{L}$.

Example 2.2. Let $\{e_j, f_j\}$ be a symplectic basis for V. If Λ_o is the lattice spanned by this basis, then it is easy to see that $\Lambda_o^* = \Lambda_o$. We say that Λ_o is *self dual.* For $1 \leq a \leq n$ (where dim $V = 2n$), let Λ_a be the lattice spanned by $\{e_j, 1 \leq j \leq n\} \cup \{f_j, a < j \leq n\} \cup \{\pi f_j, 1 \leq j \leq a\}$. Let Λ_{n+a} be the lattice spanned by $\{e_j, 1 \leq j \leq n - a\} \cup \{\pi e_j, n - a < j \leq n\} \cup \{\pi f_j, 1 \leq j \leq n\}$. Then set $\Lambda_{b+2nc} = \pi^c \Lambda_b$ for $0 \leq b < 2n$, and $c \in \mathbb{Z}$. The reader can check that the Λ_m form a complete lattice flag, which is self-dual. Precisely, $\Lambda_m^* = \Lambda_{-m}$.

We will investigate the structure of self-dual lattice flags in V. We will find in particular that any complete self-dual lattice flag is symplectically equivalent to the example just given.

The set of all lattices in V itself forms a lattice, in the sense of ordered sets, with the operations of intersection and sum as the lattice operations. We may observe that a lattice flag is trivially closed under these two operations: since its lattices are totally ordered by inclusion, the intersection of any two lattices in the flag is the smaller one, and the sum is the larger one. Lattice flags have another property: they are scalable, in the sense that if Λ is in a lattice flag, then so is $s\Lambda$ for any scalar s. To study the structure of self-dual lattice flags, we will begin by studying scalable self-dual lattices of lattices.

We will call a lattice Λ *almost self dual* if

$$\Lambda \subset \Lambda^* \subset \pi^{-1}\Lambda.$$

Proposition 2.5. *Any scalable self-dual lattice lattice contains an almost self-dual lattice.*

Proof. Let Λ be any lattice in V. We will show that by taking multiples, duals, intersections and sums, we can construct beginning with Λ an almost self-dual lattice.

We will assume the standard result from the "theory of elementary divisors", that we can find a basis $\{e_j', f_j' : 1 \leq j \leq n\}$ for Λ, such that $\langle e_j', f_\ell' \rangle = \delta_{j\ell}\pi^{c_j}$. Although the basis is not unique, the integers c_j, if arranged in decreasing order, are well-defined. We will call the c_j the *exponents* of Λ.

Given such a basis for Λ, it is then easy to see that Λ^* has a basis $\{\pi^{-c_j}e_j', \pi^{-c_j}f_j'\}$. From this, the exponents of Λ^* are seen to be $-c_j$. Also, the exponents for $\pi^d\Lambda$ are $c_j + 2d$.

We note that a lattice is almost self-dual if and only if its exponents are all either 0 or 1. If all the exponents of Λ are the same, then we can scale Λ by an appropriate power of π, to make every exponent of the rescaled lattice either 0 or 1, according as to whether the original exponent value is even or odd. The difference between the largest and smallest exponent of Λ is a measure of how far from self-dual a scalar multiple of Λ can be. We can reduce the "exponent spread" of Λ as follows. Arrange the exponents in descending order; $c_1 \geq c_2 \geq c_3 \geq \cdots \geq c_n$. Now scale Λ by the appropriate power of π to arrange that $-1 \leq c_1 + c_n \leq 2$. Now form $\Lambda \cap \Lambda^*$. It exponents will be $|c_j|$, and these will have maximum spread $\max\{|c_1|, |c_n|\}$, which is roughly half of the spread of Λ itself. It is not hard to see that by repeating this process, we can reduce the exponent spread to not more than 2, and the exponents themselves to lie in the range from 0 to 2. If the exponents are all 0 or 1, then we have an almost self-dual lattice. So we only have to worry if the values of the exponents include 2. If the lattice Λ has exponents 0, 2, and perhaps 1, we form $\Lambda' = \Lambda + \pi \Lambda^*$. We can compute that Λ' has exponents only 0 and perhaps 1. Thus Λ' is the desired almost self-dual lattice.

Now consider a self-dual lattice flag $\mathcal{L}$. If Λ is any lattice in $\mathcal{L}$, then the multiples of Λ and of Λ^* form a self-dual subflag of $\mathcal{L}$. Thus, some multiple of Λ or of Λ^* must be almost self-dual. Conversely, given an almost self dual lattice Λ, one can check that the multiples of Λ and of Λ^* do constitute a self-dual lattice flag.

Another way of formulating the fact that a self-dual lattice flag must contain an almost self-dual lattice, is to observe that every lattice Λ in the flag must have exponents of only two values, say c and $c - 1$; for as we have noted above, if the exponents of Λ are $\{c_j\}$, then the exponents of Λ^* are $\{-c_j\}$, and the exponents of $\pi^d \Lambda$ are $\{c_j + 2d\}$. If Λ is almost self-dual, then its exponents are 0 and 1, and so every multiple of Λ has exponents with only two values which differ by 1. Conversely, if Λ is a lattice with only two values for exponents, with the values differing by 1, then some multiple of Λ or of its dual will be self-dual, so that Λ does generate a self-dual lattice flag.

Now consider a self-dual lattice flag $\mathcal{L}$ which is complete. We claim that $\mathcal{L}$ must contain a self-dual lattice. We know from the discussion so far that $\mathcal{L}$ must contain at least an almost self-dual lattice. Let Λ be one. Then, since $\mathcal{L}$ is complete, it contains a lattice Λ' such that $\Lambda \subset \Lambda'$, and Λ'/Λ is a line over $\bar{k}$. The exponents of Λ' must be the same as those of Λ, except for one, which is one less than a corresponding exponent of Λ. (This can be seen, for example, by observing that the volume of Λ can be expressed in terms of its exponents.) The exponents of Λ are all either 0 or 1; if Λ is not itself self-dual, then 1 certainly occurs. Since Λ' must have exponents with only two values, it cannot have -1 as an exponent, so its exponents again must have values 0 and 1, with one more 0 and one less 1 than the exponents of Λ. Hence, we see that by looking for a maximal almost-self dual lattice in $\mathcal{L}$, we will find a lattice which is actually self-dual.

Still considering the self-dual, maximal lattice flag $\mathcal{L}$, let Λ_o be a self-dual lattice in $\mathcal{L}$. Since $\mathcal{L}$ is totally ordered by inclusion, we may label the lattices in $\mathcal{L}$ by integers, in a manner such that Λ_{m+1} is the largest element of $\mathcal{L}$ strictly contained in Λ_m. This labeling is clearly unique. Then the lattices Λ_m for $0 \leq m \leq 2n$ are contained between Λ_o and $\pi \Lambda_o = \Lambda_{2n}$. They therefore define a flag in $\overline{\Lambda} = \Lambda/\pi\Lambda$. The condition that each Λ_m should have at most two exponent values translates into the condition that the flag $\overline{U}_m = \{\Lambda_m/\pi\Lambda_o\}$ should be a self-dual flag in $\overline{\Lambda}_o$. Thus, we see that $\mathcal{L}$ is indeed isometric to the example given above of a self-dual complete lattice flag.

Iwahori-Bruhat Decomposition Now consider two complete, self-dual lattice flags. Since these are in particular lattice flags, we know that we can find a line decomposition of V which is compatible with both of them. However, as in the case of ordinary flags, we can do better.

Proposition 2.6. Iwahori-Bruhat decomposition, geometric ver–sion
For any two complete, self-dual lattice flags in V, there is a symplectic line decomposition compatible with both flags.

Proof. We label the lattices in $\mathcal{L}$ and $\mathcal{M}$ by the integers, in the standard way described above: Λ_o is the self-dual element of $\mathcal{L}$. Consecutive elements of $\mathcal{L}$ are labeled by consecutive integers, with positive integers labeling sublattices of Λ_o. The labeling of $\mathcal{M}$ is similar.

Consider the lattice Λ_o in $\mathcal{L}$. It is self-dual. The $\overline{k}$ vector space $\overline{\Lambda}_o = \Lambda_o/\pi\Lambda_o$ inherits a symplectic form by reducing modulo $\mathcal{P}$ the restriction to Λ_o of the form $\langle\ ,\ \rangle$ on V. For a lattice M_a of $\mathcal{M}$, consider the subspace $U(M_a)$ of $\overline{\Lambda}_o$ defined by $(M_a \cap \Lambda_o) + \pi\Lambda_o$.

Consider the orthogonal complement of $U(M_a)$ with respect to the symplectic form on $\overline{\Lambda}_o$. It can be lifted to a lattice contained between Λ_o and $\pi\Lambda_o$. As such, it will consist of elements $y \in \Lambda_o$ such that $\langle x, y\rangle$ lies in $\mathcal{P}$ for all x in $M_a \cap \Lambda_o$. This means that $\pi^{-1}y$ belongs to $((M_a \cap \Lambda_o) + \pi\Lambda_o)^*$. We compute that

$$((M_a \cap \Lambda_o) + \pi\Lambda_o)^* = (M_a \cap \Lambda_o)^* \cap (\pi\Lambda_o)^* = (M_a^* + \Lambda_o^*) \cap \pi^{-1}\Lambda_o$$
$$= (M_{-a} \cap \pi^{-1}\Lambda_o) + \Lambda_o = \pi^{-1}((\pi M_{-a} \cap \Lambda_o) + \pi\Lambda_o).$$

It follows that $U(M_a)^\perp = U(\pi M_{-a}) = U(M_{-a+2n})$. Here the $\perp$ is understood to refer to $\overline{\Lambda}_o$.

Thus, we see that the flag in $\overline{\Lambda}_o$ defined by $\mathcal{M}$ will be a self-dual flag. The flag defined by $\mathcal{L}$ is likewise self-dual. According to the ordinary Bruhat decomposition for $\text{Sp}(\overline{\Lambda}_o)$, we can find a symplectic basis for $\overline{\Lambda}_o$ compatible with both flags. According to the Iwahori-Bruhat decomposition for $GL(V)$, we can lift this basis to a basis of Λ_o such that the corresponding line decomposition is compatible with both lattice flags. It remains to show that we can lift the basis in such a way that we get a symplectic line decomposition - that is, we can lift the symplectic basis for $\overline{\Lambda}_o$ to a symplectic basis for Λ_o that is also compatible with both lattice flags.

As in the earlier arguments, we proceed by induction on $\dim V$. Let a_1 be the smallest index such that $M_{a_1} \subset \pi\Lambda_o$. That is, $M_{a_1} \subset \pi\Lambda_o$ but $M_{a_1-1} \not\subset \pi\Lambda_o$. Then, according to the duality computation above, $M_{-a_1} \supset \pi^{-1}\Lambda_o$, but $M_{-a_1+1} \not\supset \pi^{-1}\Lambda_o$. It follows that we can choose $e_1 \in M_{a_1-1}$ and $f_1 \in \pi M_{-a_1} \cap \Lambda_o$, such that $\langle e_1, f_1\rangle = 1$. We note that e_1 and f_1 are both in Λ_o, and that, according to the proof of the Bruhat Decomposition for vector spaces, we can choose them compatible with the flag defined by the Λ_k also. That means that, if $\overline{P}$ is the plane in $\overline{\Lambda}_o$ spanned by e_1 and f_1 (taken modulo

$\pi \Lambda_o$), then each subspace of $\overline{\Lambda}_o$ defined by Λ_k for $0 \leq k \leq 2n$, is the direct sum of its intersections with $\overline{P}$ and with $\overline{P}^\perp$. From this, it follows that, if P is the plane in V spanned by e_1 and f_1, then each lattice Λ_k satisfies $\Lambda_k = (\Lambda_k \cap P) \oplus (\Lambda_k \cap P^\perp)$. (Here of course $P^\perp$ refers to orthogonal complement in V rather than in $\overline{\Lambda}_o$.) In brief, the decomposition $V = P \oplus P^\perp$ is compatible with the lattice flag $\mathcal{L}$.

We claim that the decomposition $V = P \oplus P^\perp$ is also compatible with the lattice flag $\mathcal{M}$. To see this, consider the lattice M_o, the self-dual member of $\mathcal{M}$, and consider the flag in $\overline{M}_o$ defined by $\mathcal{M}$. We may write $a_1 - 1 = 2n\ell + b_1 - 1$ with $0 < b_1 \leq 2n$. Then $-a_1 + 2n = -2n\ell + 2n - b_1$, and $0 \leq 2n - b_1 < 2n$. With this notation, we see that $\pi^{-\ell}e_1$ is in M_o, and indeed, is in M_{b_1-1}. Similarly, $\pi^\ell f_1$ is in M_o, or more precisely, is in M_{2n-b_1}. From this, we see that the images in $\overline{M}_o$ of $\pi^{-\ell}e_1$ and $\pi^\ell f_1$ are compatible with the flag defined by the M_k. Since $\pi^{-\ell}e_1$ and $\pi^\ell f_1$ obviously span the same plane P as do e_1 and f_1, we can argue as in the case of the flag $\mathcal{L}$, that the decomposition $V = P \oplus P^\perp$ is compatible with $\mathcal{M}$, as desired.

Thanks to this compatibility, we can now finish the argument by induction on dim V, in parallel with the argument for the Bruhat Decomposition for $\mathrm{Sp}(V)$.

Parahoric Subgroups Given $\mathcal{L}$ a self-dual lattice flag in V, we denote the stabilizer in $\mathrm{Sp}(V)$ of all the lattices of $\mathcal{L}$ by $K_{\mathcal{L}}$. The groups $K_{\mathcal{L}}$ are referred to as *parahoric subgroups*. If $\mathcal{L}$ is complete, we also write $K_{\mathcal{L}} = J$. This is the *Iwahori subgroup* of $\mathrm{Sp}(V)$.

If $K \subset \mathrm{Sp}(V)$ is any compact subgroup, then as we have seen in the discussion of $\mathrm{GL}(V)$, K will preserve some lattice Λ in V. It will then also preserve all lattices obtained from L by the operations of scalar multiplication, duality, intersection and sum. In other words, it will preserve the self-dual lattice of lattices generated by L. Hence, according to Proposition 2.5, it will preserve some almost self-dual lattice, and likewise the self-dual lattice flag it generates. Hence, any compact subgroup of $\mathrm{Sp}(V)$ is contained in a parahoric subgroup.

Among the parahoric subgroups, the smallest ones are of course the stabilizers of the members of a complete self dual flag - the Iwahori subgroups. The largest are the stabilizers of a single almost self-dual lattice (and hence, of all elements of the self-dual lattice flag it generates). Unlike the case of $\mathrm{GL}(V)$, these groups are not all conjugate inside $\mathrm{Sp}(V)$. However, despite this, they all have the same Sylow p-subgroup. More precisely, we can show that the Sylow p-subgroup of the Iwahori group is the Sylow p-subgroup of any parahoric (up to conjugacy). Indeed, let H be a compact pro-p subgroup of $\mathrm{Sp}(V)$. Then H preserves some self-dual lattice flag $\mathcal{L}$. We may suppose that $\mathcal{L}$ is maximal with respect to being H-stable. We claim then that $\mathcal{L}$ is complete. If not, let $\Lambda \in \mathcal{L}$ be a maximal almost self-dual lattice in $\mathcal{L}$. If Λ is not itself self-dual, then Λ^*/Λ is a $\overline{k}$ vector space, and H acts on Λ^*/Λ via a finite quotient $\overline{H}$, which is of course a p-group. Further, Λ^*/Λ inherits a symplectic form, and $\overline{H}$ will preserve this form. Since $\overline{H}$ is a p-group, it will preserve some line $\overline{L}_1 \subset \Lambda^*/\Lambda$, and also its orthogonal subspace. The line and its orthogonal subspace lift to a nested pair of mutually dual, H-invariant lattices strictly contained between Λ and Λ^*. Also, the smaller of the two lattices will be almost self-dual. These lattices and their multiples can be adjoined to $\mathcal{L}$ to make a larger H-invariant lattice flag. Hence, maximality of $\mathcal{L}$ implies that Λ must have been self-dual. Then H acts on $\overline{\Lambda} = \Lambda/\pi\Lambda$, and will preserve a maximal, self-dual flag there, since it acts via a p-group. This flag then determines a complete self-dual lattice flag, which will evidently be invariant by H. Thus, H is contained in the pro-p part of an Iwahori subgroup, as claimed.

3 Structure of the Iwahori Hecke Algebra

3.1 $\mathcal{H}(G//K)$

First we study $\mathcal{H}(G//K)$, where $K = K(\Lambda) \subset \mathrm{GL}(V)$ is the stabilizer of a lattice in V, or, in the case of $\mathrm{Sp}(V)$, $K = K(\Lambda) \cap \mathrm{Sp}(V)$, where Λ is a self-dual lattice. In the case of GL_n, let A° be the group of all diagonal matrices whose entries are powers of π. Let $A^{\circ+}$ be the subsemigroup of A° defined by

$$A^{\circ+} = \left\{ \begin{bmatrix} \pi^{m_1} & & & \\ & \pi^{m_2} & & \\ & & \ddots & \\ & & & \pi^{m_n} \end{bmatrix} : m_1 \leq m_2 \leq \ldots \leq m_n \right\}.$$

In the case of Sp_{2n}, first choose a symplectic basis, and order it so that the matrix of the symplectic form is

$$\begin{bmatrix} 0 & -I_n \\ I_n & 0 \end{bmatrix},$$

where I_n is the $n \times n$ identity matrix. Then set

$$A^\circ = \left\{ \begin{bmatrix} \mathbf{a} & 0 \\ 0 & \mathbf{a}^{-1} \end{bmatrix} : \mathbf{a} \in \mathbf{A}^\circ \text{ for } \mathrm{GL_n} \right\},$$

and let $A^{\circ+}$ be the subsemigroup for which $\mathbf{a}$ belongs to $A^{\circ+}$ for GL_n.

The Weyl group acts on the diagonal torus by conjugation. This action permutes the diagonal entries, and it preserves the subgroup A° It is not difficult to verify that the semigroup $A^{\circ+}$ contains a unique element in each W orbit in A°. The group K contains a set of representatives for W. As noted at the end of Section 2.3, this implies that we have the decompositions $G = KA^{\circ+}K = JA^\circ K$. From the first of these decompositions, we deduce the following result.

Proposition 3.1. $\mathcal{H}(G//K)$ *is commutative.*

Proof. The proof is based on the simple and classical statement, due to Gel'fand, that if an algebra A allows an anti-automorphism which is the identity, then A is commutative. Here we start with the transpose map sending g to g^t for $g \in \mathrm{GL}_n$, which preserves K and $A^{\circ+}$. This induces an anti-automorphism of the convolution algebra $C_c^\infty(G)$ and is the identity upon restriction to $\mathcal{H}$. A similar argument works for Sp_{2n} as long as we use coordinates with respect to a symplectic basis.

Thus irreducible representations of K occur at most once in any representation of G. The operators originally constructed by Hecke were elements of $\mathcal{H}(G//K)$ where $G = \mathrm{SL}_2$ or $G = \mathrm{PGL}_2$.

3.2 $\mathcal{H}(G//J)$

The extended affine Weyl group For $G = \mathrm{GL}_n$, the Weyl group W is the group of permutations, S_n, generated by the $n - 1$ transpositions s_j for $j = 1, \ldots, n - 1$ where s_j interchanges e_j and e_{j+1}. The extended affine Weyl group, $\widetilde{W}^\circ$, is almost an affine Coxeter group generated by reflections. Generators for $\widetilde{W}^\circ$ include the $s_1, \ldots, s_{n-1}$ together with two additional generators, s_0 and t, where

$$
s_0 = \begin{bmatrix} 0 & & & \pi^{-1} \\ & 1 & & \\ & & \ddots & \\ & & & 1 & \\ \pi & & & & 0 \end{bmatrix} \qquad
t = \begin{bmatrix} 0 & 1 & & & \\ & 0 & 1 & & \\ & & \ddots & \ddots & \\ & & & & 0 & 1 \\ \pi & & & & & 0 \end{bmatrix}.
$$

The group generated by the $s_j, 1 \leq j \leq n - 1$, is the usual Weyl group of permuation matrices. If we adjoin s_0, we get a group containing all diagonal matrices of determinant one (whose diagonal entries are powers of π). Adding t gives us a group containing all diagonal matrices (whose diagonal entries are powers of π).

The choice of t is based on the fact that $ts_jt^{-1} = s_{j-1}$ so t normalizes J_{GL_n}, where

$$
J_{\mathrm{GL}_n} = \begin{bmatrix} \mathcal{O}^\times & \mathcal{O} & \ldots & \mathcal{O} \\ \mathcal{P} & \ddots & \ddots & \vdots \\ \vdots & \ddots & & \mathcal{O} \\ \mathcal{P} & \ldots & \mathcal{P} & \mathcal{O}^\times \end{bmatrix}.
$$

The relations in $\widetilde{W}^\circ$ include the usual Coxeter relations, $s_i^2 = 1$ (but now with $0 \leq i \leq n - 1$) and $(s_i s_j)^{m_{ij}} = 1$, where for GL_n, $m_{i,i+1} = 3$ and

$m_{ij} = 2$ if $|i - j| > 1$ with $|i - j|$ computed modulo n. But we also have the relations $t s_j t^{-1} = s_{j-1}$.

For $G = \mathrm{Sp}_{2n}$, $J_{\mathrm{Sp}_{2n}} = J_{\mathrm{GL}_n} \cap \mathrm{Sp}_{2n}$ and the generators of the group $\widetilde{W}^{\circ}$ can most easily be described by listing only the symplectic basis vectors that they change in some way (here $1 \le j \le n - 1$):

$$s_0 : e_1 \mapsto \pi f_1 \qquad s_j : e_j \leftrightarrow e_{j+1} \qquad s_n : e_n \mapsto f_n$$
$$f_1 \mapsto -\pi^{-1} e_1 \qquad f_j \leftrightarrow f_{j+1} \qquad f_n \mapsto -e_n.$$

For both of these versions of $\widetilde{W}^{\circ}$, the length of a word is defined to be the minimum number of generators of type s_j (i.e. excluding any occurrences of the generator t in the case of GL_n). This is used so that $\mu(JwJ) = q^{\ell(w)}$.

Iwahori-Matsumoto presentation The relations for $\mathcal{H}(G//J)$ consist of two types. To describe the first, in place of $(s_i s_j)^{m_{ij}} = 1$, we use the braid relation $s_i s_j s_i \cdots = s_j s_i s_j \cdots$, where each side is a reduced expression for the longest word in W. This avoids the use of inverses. Since

$$l(s_i s_j s_i \cdots) = l(s_j s_i s_j \cdots) = m_{ij} \text{ and}$$
$$l(uv) = l(u) + l(v) \text{ implies } \mu(JuvJ) = \mu(JuJ)\mu(JvJ),$$

we know

$$\mu(J s_i s_j s_i \cdots J) = \mu(J s_i J)\mu(J s_j J)\mu(J s_i J) \cdots = q^{m_{ij}} \text{ and}$$
$$\mu(J s_j s_i s_j \cdots J) = \mu(J s_j J)\mu(J s_i J)\mu(J s_j J) \cdots = q^{m_{ij}}.$$

Thus the braid relations hold in $\mathcal{H}(G//J)$:

$$f_{s_i} f_{s_j} f_{s_i} \cdots = f_{s_i s_j s_i \cdots} = f_{s_j s_i s_j \cdots} = f_{s_j} f_{s_i} f_{s_j} \cdots .$$

Secondly, the quadratic relations $s_i^2 = 1$ become $f_{s_i}^2 = (q-1)f_{s_i} + q f_1$.

Theorem 3.1. $\mathcal{H}(G//J)$ *is the algebra generated by the appropriately indexed* f_{s_i} *(and* f_t *when* $G = \mathrm{GL}_n$*) subject to the following relations.*

1. $f_{s_i}^2 = (q - 1)f_{s_i} + q f_1$
2. $f_u f_v = f_{uv}$ *if* $\ell(u) + \ell(v) = \ell(uv)$
3. $f_t f_{s_i} = f_{s_{i+1}} f_t$ *when* $G = \mathrm{GL}_n$

Notice that the third relation uses that $f_{t^{-1}} = f_t^{-1}$, which follows from the fact that t is by definition of length zero.

This presentation shows that the structure of $\mathcal{H}(G//J)$ is similar to that of a Coxeter group and allows us to view $\mathcal{H}$ as a deformation of $\widetilde{W}^{\circ}$ where the operators have eigenvalues q and -1 rather than 1 and -1. However, it obscures the abelian subgroup of diagonal matrices inside $\widetilde{W}^{\circ}$, which is often useful. This abelian subgroup is revealed by an anternative presentation developed by J. Bernstein and A. Zelevinsky.

Conversion to the Bernstein-Zelevinsky presentation We now show how, in the case of GL(V), to convert from the Iwahori-Matsumoto presentation given above to the Bernstein-Zelevinsky presentation. This presentation is often used, in part because it makes clear the presence of a large abelian subalgebra inside $\mathcal{H}(G//J)$ corresponding to the diagonal subgroup of $\widetilde{W}^\circ$.

In the affine Weyl group $\widetilde{W}^\circ$ for GL$_n$, set

$$a_k = \begin{bmatrix} \pi^{-1} & & & & & & \\ & \ddots & & & & & \\ & & \pi^{-1} & & & & \\ & & & 1 & & & \\ & & & & \ddots & & \\ & & & & & 1 \end{bmatrix},$$

where the first k entries along the diagonal are π^{-1}. These elements with $1 \leq k \leq n$ generate a free semigroup of rank n inside $\widetilde{W}^\circ$ and it is easy to check that $\ell(a_k) = k(n-k)$. Notice that

$$s_k a_k s_k = \begin{bmatrix} \pi^{-1} & & & & & & & \\ & \ddots & & & & & & \\ & & \pi^{-1} & & & & & \\ & & & 1 & & & & \\ & & & & \pi^{-1} & & & \\ & & & & & 1 & & \\ & & & & & & \ddots & \\ & & & & & & & 1 \end{bmatrix} \quad \text{and}$$

$$a_k s_k a_k s_k = \begin{bmatrix} \pi^{-2} & & & & & & & \\ & \ddots & & & & & & \\ & & \pi^{-2} & & & & & \\ & & & \pi^{-1} & & & & \\ & & & & \pi^{-1} & & & \\ & & & & & 1 & & \\ & & & & & & \ddots & \\ & & & & & & & 1 \end{bmatrix} = a_{k-1} a_{k+1},$$

where each of these has the first $k-1$ entries equal to π^{-2}.

Note also that s_k commutes with a_j for $j \neq k$ and therefore in particular with a_{k-1} and a_{k+1}. If we rewrite the above identity as $(a_k s_k) a_k = a_{k-1} a_{k+1} s_k$, then the length of the element of $\widetilde{W}^\circ$ in the equation is the sum of the lengths of the factors, so the identity

$$f_{a_k s_k} f_{a_k} = f_{a_{k-1}} f_{a_{k+1}} f_{s_k}, \tag{4}$$

is valid in the Hecke algebra $\mathcal{H}(\mathrm{GL}_n//J)$. Since $\ell(a_k s_k) = \ell(a_k) - 1$, the identity $f_{a_k} = f_{a_k s_k} f_{s_k}$ also holds in $\mathcal{H}(\mathrm{GL}_n//J)$. Now we use this together with the quadratic relation to see that

$$f_{a_k} f_{s_k} = f_{a_k s_k} f_{s_k}^2 = f_{a_k s_k}((q-1) f_{s_k} + q)$$
$$= (q-1) f_{a_k s_k} f_{s_k} + q f_{a_k s_k} = (q-1) f_{a_k} + q f_{a_k s_k}.$$

Solving this for $f_{a_k s_k}$ and substituting into the identity 4 yields

$$\tfrac{1}{q} f_{a_k} f_{s_k} f_{a_k} - \left(1 - \tfrac{1}{q}\right) f_{a_k}^2 = f_{a_{k-1}} f_{a_{k+1}} f_{s_k}.$$

Since the elements on the right commute, reorder them to $f_{a_{k+1}} f_{s_k} f_{a_{k-1}}$ and then multiply throughout by $f_{a_k}^{-1}$ on the left and $f_{a_{k-1}}^{-1}$ on the right. (Notice that $f_{a_{k-1}}$ is indeed invertible. More generally, for any w in the affine Weyl group, f_w is invertible, since it is a product of generators of $\mathcal{H}(G//J)$, and the generators are obviously invertible, by the quadratic relations.) This yields

$$\tfrac{1}{q} f_{s_k}(f_{a_k} f_{a_{k-1}}^{-1}) - \left(1 - \tfrac{1}{q}\right)(f_{a_k} f_{a_{k-1}}^{-1}) = f_{a_k}^{-1} f_{a_{k+1}} f_{s_k} = f_{a_{k+1}} f_{a_k}^{-1} f_{s_k}. \quad (5)$$

In equation 5, replace f_w by $q^{\frac{-\ell(w)}{2}} f_w$ and rename by setting

$$T_k = q^{-\frac{1}{2}} f_{s_k}, \quad y_k = q^{-\frac{n-2k+1}{2}} f_{a_k} f_{a_{k-1}}^{-1}, \quad \text{and } y_{k+1} = q^{-\frac{n-2k-1}{2}} f_{a_{k+1}} f_{a_k}^{-1}.$$

Then 5 becomes

$$\tfrac{1}{q} q^{\frac{1}{2}} T_k q^{\frac{n-2k+1}{2}} y_k - \left(\tfrac{q-1}{q}\right) q^{\frac{n-2k+1}{2}} y_k = q^{\frac{n-2k-1}{2}} y_{k+1} q^{\frac{1}{2}} T_k.$$

Combining powers of q, dividing by $q^{\frac{n-2k}{2}}$, and rearranging produces

$$T_k y_k - y_{k+1} T_k = (q^{\frac{1}{2}} - q^{-\frac{1}{2}}) y_k.$$

Using the action $s_k(y_k) = y_{k+1}$, this can be rewritten in the form of the Bernstein-Zelevinsky relation

$$T_k y_k - s_k(y_k) T_k = (q^{\frac{1}{2}} - q^{-\frac{1}{2}}) \frac{s_k(y_k) - y_k}{s_k(y_k) y_k^{-1} - 1}. \quad (6)$$

Note also that

$$T_k y_j = y_j T_k \text{ for } j \neq k, k+1$$

since then s_k commutes with a_j and a_{j-1} as observed above. In addition, the y_j generate an abelian subalgebra and the T_k satisfy the relations

$$T_i T_j = T_j T_i \qquad \text{if } |i - j| > 1$$
$$T_k T_{k+1} T_k = T_{k+1} T_k T_{k+1} \qquad \text{for } 1 \leq k \leq n - 1.$$

Also, the relation in 6 in fact holds for any Laurent polynomial in the y_j.

4 Results on representations of G

4.1 Fundamental techniques

For more details, see the references for further reading.

Parabolic induction The primary method of constructing representations of p-adic groups is parabolic induction. Begin with a parabolic subgroup P of the group G and decompose it as $P = M_P U_P$ where M_P is a Levi component and U_P is the unipotent radical of P. Let σ be a (smooth admissible) representation of M_P on the space Y and lift σ to P by taking it to be trivial on U_P. The induced representation $\mathrm{Ind}_P^G(\sigma)$, called a *principal series*, consists of the space

$$\{f : G \to Y : f(pg) = \sigma(p)f(g) \text{ for all } p \in P, g \in G\}$$

and the action

$$\mathrm{Ind}_P^G(\sigma)(g)f(h) = f(hg).$$

It is possible to prove the following facts.

1. If σ is admissible then $\mathrm{Ind}_P^G(\sigma)$ is admissible.
2. $(\mathrm{Ind}_P^G \sigma)^* = \mathrm{Ind}_P^G(\delta_P \sigma^*)$ where δ_P is the modular function of P defined using p-adic absolute value by $\delta_P(m) = |\det \mathrm{Ad}(m|_{\mathfrak{u}_P})|_p^{-1}$
3. If σ is unitary then $\mathrm{Ind}_P^G(\delta_P^{1/2}\sigma)$ is unitary.

In the special case when the parabolic subgroup is the Borel subgroup, then the Levi component is a torus and the decomposition is $P = B = AU$. Since A is abelian, take ψ to be a character trivial on the compact part of the torus, i.e. a character of $A/A_0 \simeq A^0$ where $A_0 = A \cap K$. Then the induced representation $\mathrm{Ind}_B^G(\psi)$ is called an *unramified minimal principal series*, or a *spherical principal series*. Since $G = KB$, each spherical principal series contains a unique K-fixed vector. (In fact we will see below that the process of forming spherical principal series will yield all representations with K-fixed vectors.) These induced representations are highly important in the study of automorphic forms since any given representation of an adele group over a global field factors into a tensor product of representations of groups over local fields, and almost all factors are spherical principal series.

Jacquet modules Let ρ be a smooth (admissible) representation of G. If ρ can be realized as a principal series from some parabolic, then it must be induced from a representation trivial on U_P, the unipotent radical of the parabolic. So we look for invariants for U_P. Set $Y(U_P) = \{\rho(u)y - y : y \in Y, u \in U_P\}$. This records all nontrivial action of U_P. The quotient $Y_{U_P} = Y/Y(U_P)$ is the maximal quotient of Y on which U_P acts trivially. It is called a *Jacquet module*.

Proposition 4.1. *$Y(U_P)$ is an M_P-submodule of Y.*

The proof involves a simple calculation. We denote the action of M_P on Y_{U_P} by ρ_U.

Frobenius Reciprocity

Proposition 4.2. *Let σ be a smooth representation of M and let ρ be a smooth representation of G. Then*

$$\operatorname{Hom}_G(\rho, \operatorname{Ind}_P^G(\sigma)) \simeq \operatorname{Hom}_{M_P}(\rho_U, \sigma).$$

So if $Y_{U_P} \neq \{0\}$ then ρ is a constituent of the principal series representation arising from P. Let J_U denote the quotient map from Y (admissible as a G-module, but not as an M-module) to Y_{U_P} (which is admissible as an M-module).

Proposition 4.3. *Let U_P^- be the unipotent subgroup of G opposite to U_P. Suppose H is a compact subgroup of G that can be factored as $H = (H \cap U_P^-)(H \cap M)(H \cap U_P)$. Then $J_U : Y^H \to Y_{U_P}^{H \cap M_P}$ is surjective.*

Notice that such a factorization is often possible, for example whenever H is a principal congruence subgroup. In fact, for all sufficiently small compact subgroups of M it is possible to find an H as in the proposition. Then when Y^H is finite dimensional, $Y_{U_P}^{H \cap M_P}$ will also be finite dimensional and thus if ρ is admissible, so is ρ_{U_P}. This means that it is possible to check whether a representation occurs as part of a principal series by checking if some Jacquet module is nonzero. If all Jacquet modules are zero (i.e. $Y_{U_P}(\rho) = 0$ for all P) then we say ρ is *cuspidal*, meaning it has compactly supported matrix coefficients. In particular, they are square-integrable (a.k.a. discrete series). Such representations are characteristic of p-adic group representation theory.

Harish-Chandra philosophy of cusp forms The above background leads to the philosophy of cusp forms, which involves a two-step program to classify representations of G.

1. Classify cuspidal representations.
2. Decompose all principal series $\operatorname{Ind}_P^G(\sigma)$ where σ is cuspidal.

Completing this program will guarantee that all representations have been found, but will involve redundancy since a representation could appear in more than one induced representation. However, such cases are very limited, as demonstrated in the following proposition. First, notice that for fixed M, it is possible to choose U and $P = MU$ so that ρ appears as a submodule of $\operatorname{Ind}_P^G(\sigma)$ since J_U is exact. Thus by $\rho \in \operatorname{Ind}_P^G(\sigma)$ it is meant that ρ is a submodule of $\operatorname{Ind}_P^G(\sigma)$.

Proposition 4.4. *For $\rho \in \widehat{G}$, if $\rho \in \operatorname{Ind}_P^G(\sigma)$ and $\rho \in \operatorname{Ind}_{P'}^G(\sigma')$, then (M_P, σ) is conjugate to $(M_{P'}, \sigma')$, meaning that M_P is conjugate to $M_{P'}$ in such a way that σ is sent to σ'.*

4.2 Connection of Jacquet functor to $\mathcal{H}(G//J)$

Recall the semigroup $A^{\circ +}$ introduced in Section 3.1. This subsemigroup of the torus contains one representative of each conjugacy class in A°. Under conjugation, $A^{\circ +}$ stretches U_P^+ and shrinks U_P^- (upper and lower triangular matrices respectively) in the sense that for $a \in A^{\circ +}$,

$$a(J \cap U_P^+)a^{-1} \supset J \cup U_P^+ \text{ and}$$
$$a(J \cap U_P^-)a^{-1} \subset J \cup U_P^-.$$

This implies that $\ell(ab) = \ell(a) + \ell(b)$ for all $a, b \in A^{\circ +}$ and therefore that $f_a * f_b = f_{ab}$. Thus $\{f_a : a \in A^{\circ +}\} \simeq A^{\circ +}$ is a semigroup.

Let H be a subgroup of G satisfying the properties in Proposition 4.3 and properties analogous to those above with respect to $A^{\circ +}$. Then for $f_a = \frac{1}{\mu(H)}\chi_{HaH}$ (normalized to be idempotent when $a = 1$), $J_U(\rho(f_a)y) = \delta_P(a)\rho_{U_P}(f_a^M)J_U(y)$, where $f_a^M = \frac{1}{\mu_M(H \cap M)}\chi_{(H \cap M)a(H \cap M)}$. Thus the action of an abelian subsemigroup in $\mathcal{H}(G//J)$ transfers to an action of a subsemigroup corresponding to M. Since the f_a form a commutative semigroup, we can write

$$Y^H = \sum_{\lambda \in \widehat{A^{\circ +}}} Y^{H,\lambda}$$

where the $Y^{H,\lambda}$ are generalized eigenspaces with $\lambda = (\lambda(a_1), \lambda(a_2), \dots)$ for the canonical generators a_k defined as in 3.2.

Corollary 4.1. J_U maps $Y^{H,\lambda}$ to $Y_{U_P}^{H \cap M, \delta_P \lambda}$ and is surjective.

The proof involves considering the difference between G and M in terms of $A^{\circ +}$.

If elements of M are written as block diagonal matrices and the index k ends a block, then a_k becomes central in M and therefore $f_{a_k}^M$ must be invertible. So J_U acting on $Y^{H,\lambda}$ is either an isomorphism or zero, depending on whether $\lambda(a_k) \neq 0$ or $\lambda(a_k) = 0$, for a_k in the center of M.

4.3 Category equivalence

The consequence for $\mathcal{H}(G//J)$ is the following category equivalence. First, recall that for $H = J$, all f_a are invertible, as noted in Section 3.2. Therefore for the generalized eigenspace $Y^{J,\lambda}$, there are no zero components in λ. This implies $J_{U_B}(Y^J)$ is injective and therefore an isomorphism.

Corollary 4.2 (Borel).

1. If $\rho \in \widehat{G}$, then $\rho^J \neq \{0\}$ if and only if ρ is a submodule of the principal series.

2. $\mathcal{M}(G, e_J) \simeq \mathcal{M}(\mathcal{H}(G//J))$ where the first is the category of G-modules generated by J-fixed vectors and the second is the category of $\mathcal{H}(G//J)$-modules.

Proof. 1. Given a vector v in ρ^J, it is sent by the Jacquet functor J_U to a vector that transforms according to an admissible representation of A. Hence it can be decomposed as a sum of generalized eigenvectors for A, with each generalized eigenvector associated to a character of A. Since $J_U(v)$ will be $A \cap J = A_0$ invariant, all those characters must be trivial on A_0, i.e., unramified. Thus, the associated induced representations, into which ρ embeds by Proposition 4.2, are spherical principal series.

2. As noted in Section 1, the main point is to show that $\mathcal{M}(G, e_J)$ is closed under taking submodules. Suppose Y is generated by Y^J, but $Z \subset Y$ has $Z^J = \{0\}$. Then Z can be mapped non-trivially into some $\mathrm{Ind}_P^G(\sigma)$, for appropriate parabolic P and supercuspidal representation σ. This mapping is found by first taking the Jacquet functor to P, and then projecting to the component that transforms according to σ, and finally using Frobenius Reciprocity (Proposition 4.2). Since J_U is exact, and likewise projection to a supercuspidal component, there must also be a non-trivial map from Y to $\mathrm{Ind}_P^G(\sigma)$. The fact that Y is generated by Y^J guarantees that $(\mathrm{Ind}_P^G \sigma)^J \neq \{0\}$. According to the remark at the beginning of this section, this means that σ must have a non-trivial Jacquet module with respect to a Borel subgroup, and with respect to unramified characters of the maximal torus. According to Proposition 4.4, this means that P is already a Borel, and σ an unramified character - that is, Z maps to a spherical principal series, contrary to the assumption that it did not have any J-fixed vectors.

Thus $\mathcal{H}(G//J)$ provides a precise understanding of the spherical principal series of the group G, in other words about the algebraic structure of G. We have not addressed the unitary structure or the Plancherel formula. It is in fact also possible to detect which representations of G are unitary by use of $\mathcal{H}(G//J)$. This is due to Barbasch and Moy.

5 Spherical Function Algebras

There are many other representations of G besides spherical principal series, and although the general representation is far from completely known, we will now indicate how to begin to extend the above result. We will look especially at the minimal principal series, that is, representations induced from an arbitrary character of the Borel subgroup.

As a first strategy one might think to look at algebras $\mathcal{H}(G//K)$ as K is made smaller and smaller. However these algebras become more and more complicated and there is a great deal of redundancy since any J-fixed vector is also a K-fixed vector for $K \subset J$. We would like to filter out those representations we already know and understand. A way of doing this is to look at non-trivial representations of compact groups. This leads to the study of spherical function algebras.

5.1 Structure

Let K be an open compact subgroup of a locally compact group G and let $\rho : K \to \mathrm{GL}(V)$ be an irreducible representation of K. $C_c^\infty(G; \mathrm{End}(V))$ is the algebra of matrix valued functions on G under convolution and is isomorphic to $C_c^\infty(G) \otimes \mathrm{End}(V)$. Define the following subalgebra of $C_c^\infty(G; \mathrm{End}(V))$:

$$\mathcal{H}(G//K; \rho) = \{f : G \to \mathrm{End}(V) \mid f(k_1 g k_2) = \rho(k_1) f(g) \rho(k_2)\}.$$

This is called a *spherical function algebra,* or also, a Hecke algebra. Normalize and extend ρ in the following way,

$$e_\rho(g) = \begin{cases} \rho(k)/\mu_G(K) & g = k \in K \\ 0 & g \notin K. \end{cases}$$

Then e_ρ is idempotent as an element of $C_c^\infty(G; \mathrm{End}(V))$ and we have

$$\mathcal{H}(G//K; \rho) = e_\rho * C_c^\infty(G; \mathrm{End}(V)) * e_\rho.$$

Just as $\mathcal{H}(G//K)$ determines the irreducible representations of G which contain K-fixed vectors, $\mathcal{H}(G//K; \rho)$ describes those irreducible representations of G which contain ρ^* on restriction to K.

Support issues Although double coset algebras have a natural basis consisting of the χ_{KgK}, the situation is more complicated with spherical function algebras, both because there may be more than one element supported on a given coset, and because some cosets may support no elements of the algebra. In order to characterize those cosets which do support an element of $\mathcal{H}(G//K; \rho)$, first notice that the mapping $f \to f(g)$ defines an isomorphism

$$\mathcal{H}(G//K; \rho)|_{KgK} \simeq \mathrm{Hom}_{K_g}(\rho, \mathrm{Ad}^* g(\rho)),$$

where

$$K_g = K \cap gKg^{-1},$$
$$\mathrm{Ad}g(K) = gKg^{-1},$$

and for $h \in \mathrm{Ad}g(K)$,

$$\mathrm{Ad}^* g(\rho)(h) = \rho(g^{-1}hg).$$

Elements of $\text{Hom}_{K_g}(\rho, \text{Ad}^* g(\rho))$ are intertwining maps between the two representations, and there exists an element in $\mathcal{H}(G//K; \rho)$ with support on KgK if there exist nontrivial intertwining operators, i.e. if $\text{Hom}_{K_g}(\rho, \text{Ad}^* g(\rho)) \neq \{0\}$. We say that g *intertwines* ρ, or g *is in the support of* $\mathcal{H}(G//K; \rho)$. More generally, suppose G is a locally compact group and K_1 and K_2 are two compact open subgroups of G. Given irreducible representations ρ_j of K_j, on spaces V_j, $j = 1, 2$, we say that $g \in G$ *intertwines* ρ_1 and ρ_2 if there exists a non-zero T in $\text{Hom}(V_2, V_1)$, such that $\rho_1(k)T = T\rho_2(g^{-1}kg)$, where $k \in K_1 \cap gK_2g^{-1}$.

5.2 Lie lattices and characters

The simplest way to determine whether g is in the support of certain spherical function algebras is to use the exponential map to reduce the computation to looking at the Lie algebra of G. Suppose G is a p-adic group with p sufficiently large and suppose G has Lie algebra $\mathfrak{g} \subset \text{End}(X)$.

Recall that $\mathcal{P}$ is the maximal ideal of the ring $\mathcal{O}$ of integers in our base field k. Let $\mathcal{P}^j$, for $j \in \mathbb{Z}$ be the j-th power of $\mathcal{P}$—it is the set of all elements $z\pi^j$, with $z \in \mathcal{O}$. We define the valuation function ord_k on k by setting

$$\text{ord}_k(x) = j \quad \text{if } x \in \mathcal{P}^j - \mathcal{P}^{j+1} = \pi^j \mathcal{O}^\times.$$

Observe that $\text{ord}_k(xy) = \text{ord}_k(x) + \text{ord}_k(y)$. A sequence $\{x_n\}$ in k will converge to zero if and only if $\text{ord}_k(x_n)$ goes to $+\infty$.

Consider the exponential mapping, defined by the usual power series:

$$\exp(x) = \sum_{n=0}^{\infty} \frac{x^n}{n!}.$$

This will converge if and only if the individual terms go to zero, which means that their valuations should go to ∞. We have $\text{ord}_k\left(\frac{x^n}{n!}\right) = n\,\text{ord}_k(x) - \text{ord}_k(n!)$. We can calculate that

$$\text{ord}_k(n!) = \sum_{a=1}^{n} \text{ord}_k(a) = \text{ord}_k(p)\left(\left[\frac{n}{p}\right] + \left[\frac{n}{p^2}\right] + \left[\frac{n}{p^3}\right] + \ldots\right)$$
$$\leq \text{ord}_k(p)\frac{n}{p-1}.$$

Hence $\text{ord}_k\left(\frac{x^n}{n!}\right) \geq n\left(\text{ord}_k(x) - \frac{\text{ord}_k(p)}{p-1}\right)$. Thus $\exp(x)$ makes sense providing that $\text{ord}_k(x) > \frac{\text{ord}_k(p)}{p-1}$. In particular, if p is sufficiently large relative to the degree of k over the p-adic numbers $\mathbb{Q}_p$, the function $\exp$ is defined on the whole prime ideal $\mathcal{P}$.

Set

$$
J = \begin{bmatrix} \mathcal{P} & \mathcal{O} & \dots & \mathcal{O} \\ \vdots & \ddots & \ddots & \vdots \\ & & & \mathcal{O} \\ \mathcal{P} & \dots & & \mathcal{P} \end{bmatrix}.
$$

If p is sufficiently large and the dimension of the base field over $\mathbb{Q}_p$ can be bounded, then the exponential map is defined on J and $\exp(J)$ is a maximal pro-p subgroup of $\mathrm{GL}(X)$. Consider a *Lie lattice*, Λ, inside J, i.e. a lattice such that $[\Lambda, \Lambda] \subset \Lambda$. The exponential of a Lie lattice need not be a group, but a mild condition will guarantee that it is.

Definition 5.1. *A lattice Λ is* elementarily exponentiable *or* e.e. *if* $\mathrm{ad}^{p-1}(y)(\Lambda) \subset p^2\Lambda$ *for all* $y \in \Lambda$.

If the Lie lattice Λ is elementarily exponentiable, then $\exp(\Lambda)$ is a group and therefore acts on the Lie algebra $\mathfrak{g}$ via the adjoint action. Furthermore, given a character $\psi \in \hat{\Lambda}$,

$$
\mathrm{Ad}(\exp \Lambda)(\psi) = \psi \quad \text{if and only if} \quad [\Lambda, \Lambda] \subset \ker \psi.
$$

Either of these equivalent conditions then implies that $\psi \circ \log$ is a character of $\exp \Lambda$, so this provides a useful way to construct characters.

For semisimple groups these results generalize as follows. Let $\mathfrak{g}$ be a semisimple Lie algebra with Killing form κ. Suppose Θ is an additive character of the base field k such that $\mathcal{O} \subset \mathrm{Ker}\Theta$ and $\pi^{-1}\mathcal{O}$ is not contained in $\ker \Theta$. Given a lattice $\Lambda \subset \mathfrak{g}$ with $\bar{\Lambda} \simeq \mathfrak{g}/\Lambda^\perp$, where $\Lambda^\perp$ is the annihilator of Λ with respect to κ, define $\psi_x(y) = \Theta(\kappa(x,y))$ for $x \in \mathfrak{g}$ and $y \in \Lambda$. (Note that it is not necessary to use the Killing form here. Any invariant bilinear form with no factors of p in its normalization would work.)

Proposition 5.1. *Given $g \in G$ and two e.e. lattices Λ_1 and Λ_2 in $\mathfrak{g}$ with respective characters ψ_{z_1} and ψ_{z_2}, g intertwines $\psi_{z_1} \circ \log$ on $\exp \Lambda_1$ with $\psi_{z_2} \circ \log$ on $\exp \Lambda_2$ if and only if $(z_1 + \Lambda_1^\perp) \cap \mathrm{Ad}g(z_2 + \Lambda_2^\perp) \neq \emptyset$.*

(Here we assume $\psi_{z_j} \circ \log$ is a character on $\exp \Lambda_j$ for $j = 1, 2$.) This proposition provides a geometric condition for intertwining.

Kirillov bicharacter Let Λ be an e.e. lattice. For $x, y \in \Lambda$, define

$$
\psi_z([x,y]) = \Theta(\kappa(z, [x,y])) = \Theta(\kappa([z,x], y)) = B_{\psi_z}(x,y).
$$

This is the Kirillov bicharacter, a skew symmetric bilinear form that defines a character in x and y. Then $\psi_z \circ \log$ is a character if the bicharacter is trivial, i.e. if $\psi_z([x,y]) = 1$.

Example 5.1 ($\mathfrak{sl}_2$). Choose a basis $\{h, e^+, e^-\}$ for $\mathfrak{sl}_2$ such that

$$[h, e^+] = 2e^+$$
$$[h, e^-] = -2e^-$$
$$[e^+, e^-] = h.$$

If Λ_o equals the usual Cartan decomposition,

$$\Lambda_o = \mathfrak{t}_o \oplus \mathfrak{g}_{\alpha,o} \oplus \mathfrak{g}_{-\alpha,o} = \mathcal{O}h \oplus \mathcal{O}e^+ \oplus \mathcal{O}e^-,$$

then set $\Lambda_\ell = \pi^\ell \Lambda_o$. Take $\ell > 1$ and $\psi = \psi_{ah}$ where $a \in \pi^{-\ell}\mathcal{O}^\times$.

We now compute the Kirillov bicharacter. There is no contribution to B_ψ except from commutators of elements in $\mathfrak{g}_{\alpha,o}$ and $\mathfrak{g}_{-\alpha,o}$ since each of the subalgebras in the Cartan decomposition is commutative. Thus it suffices to compute, for $b, c \in \mathcal{O}$,

$$B_\psi(be^+, ce^-) = \Theta(\kappa(ah, [be^+, ce^-]))$$
$$= \Theta(\kappa([ah, be^+], ce^-))$$
$$= \Theta(2abc\kappa(e^+, e^-)) = \Theta(8abc).$$

Set $\Lambda_{k,r,s} = \mathfrak{t}_k \oplus \mathfrak{g}_{\alpha,r} \oplus \mathfrak{g}_{-\alpha,s}$ where $\mathfrak{t}_k = \pi^k\mathfrak{t}_o$, $\mathfrak{g}_{\alpha,r} = \pi^r\mathfrak{g}_{\alpha,o}$ and $\mathfrak{g}_{-\alpha,s} = \pi^s\mathfrak{g}_{-\alpha,o}$. This is a lattice in $\mathfrak{sl}_2$. Then $B_\psi = 1$ on $\Lambda_{k,r,s}$ if and only if $r + s \geq \ell$. Take $r + s = \ell$ so that $\tilde{\psi} = \psi \circ \log$ is a character. Set $L = L_{k,r,s} = \exp(\Lambda_{k,r,s})$ and take $k < \ell$ so that B_ψ is not trivial on all of $\Lambda_{k,r,s}$.

5.3 Harish-Chandra homomorphism for SL$_2$

Let $\tilde{\psi}$ and L be as above. Let

$$A = \left\{ \begin{bmatrix} b & 0 \\ 0 & b^{-1} \end{bmatrix} : b \in k^\times \right\}$$

be the diagonal torus in SL$_2$. Here and below, we will denote the matrix $\begin{bmatrix} b & 0 \\ 0 & b^{-1} \end{bmatrix}$ simply by the element b of $k^\times$ appearing in its $(1,1)$-entry. For each b in A, it is easy to see that there is a function f_b of the spherical function algebra $\mathcal{H}(\mathrm{SL}_2 // L, \tilde{\psi})$ such that f_b is supported on LbL, and $f_b(\ell_1 b \ell_2) = \tilde{\psi}(\ell_1 \ell_2)$.

Proposition 5.2. *1. Supp*$(\mathcal{H}(\mathrm{SL}_2 // L, \tilde{\psi})) = LAL$.
2. Setting $\tilde{f}_b = f_b|_A$, the normalized restriction map

$$\tau : f_b \to \frac{(\mu_{\mathrm{SL}_2}(L)\mu_{\mathrm{SL}_2}(LBL))^{1/2}}{\mu_A(L \cap A)} \tilde{f}_b$$

is an isomorphism between $\mathcal{H}(\mathrm{SL}_2 // L, \tilde{\psi})$ and $\mathcal{H}(A // (L \cap A)), \tilde{\psi})$.

Remark 5.1.

The Hecke algebra $\mathcal{H}(A//(L\cap A),\tilde{\psi})$ is rather easy to understand. Note that, since A is commutative, double cosets are equal to left or right cosets. Thus, $(L\cap A)b(L\cap A) = b(L\cap A)$. Denote by $\hat{f}_b$ the characteristic function of $b(L\cap A)$. Since $\tilde{\psi}$ can be extended from $L\cap A$ to a character of all of A, the map $\hat{f}_b \to \hat{f}_b$ defines an isomorphism from $\mathcal{H}(A//(L\cap A),\tilde{\psi})$ to $C_c^\infty(A/(L\cap A))$, which is just the convolution algebra of a discrete abelian group.

The map τ is in fact an L^2 isometry, up to the factor $\frac{\mu_{\mathrm{SL}_2}(L)}{\mu_A(L\cap A)}$. This means that the Plancherel measure of the series of representations of SL_2 associated to $\mathcal{H}(A//(L\cap A),\tilde{\psi})$ is essentially the same as the Plancherel measure for $A/(L\cap A)$ (which is just Haar measure on the Pontrjagin dual of $A/(L\cap A)$).

Proof (Proof of Proposition 5.2). For simplicity in the following argument, we will assume that Haar measures have been normalized so that $\mu_{\mathrm{SL}_2}(L) = 1 = \mu_A(L\cap A)$. This does not affect the argument in any essential way, but does simplfy some formulas, in which factors involving these two measures would have to appear, absent such a normalization.

We first assume statement (i) and prove statement (ii). Write $A = A^+ \cup A^-$, where

$$A^+ = \left\{ \begin{bmatrix} b & 0 \\ 0 & b^{-1} \end{bmatrix} : b^{-1} \in \mathcal{O} \right\}$$

and

$$A^- = \left\{ \begin{bmatrix} b & 0 \\ 0 & b^{-1} \end{bmatrix} : b \in \mathcal{O} \right\}$$

If we write $b = \pi^\ell u$, where u is a unit in $\mathcal{O}$, we have that $Ad(b)(\Lambda_{k,r,s}) = \Lambda_{k,r+2\ell,s-2\ell}$. Thus A^+ is the semigroup of elements of A that stretch $\mathfrak{g}_\alpha$, and shrink $\mathfrak{g}_{-\alpha}$, and A^- is the semigroup of elements that do the opposite. In particular, if b_1 and b_2 are any two elements of A^+, then

$$\mu_{\mathrm{SL}_2}(Lb_1L)\mu_{\mathrm{SL}_2}(Lb_2L) = \mu_{\mathrm{SL}_2}(Lb_1b_2L). \tag{7}$$

Given this, a very general argument lets us show that $f_{b_1} * f_{b_2} = f_{b_1b_2}$. Indeed, let $Lb_1L = \cup_j \ell_j b_1 L$, and $Lb_2L = \cup_k \ell'_k b_2 L$ be decompositions of the respective double cosets of L into right cosets. We may assume that one of the ℓ_j is the identity, and likewise for the ℓ'_k. Then, under the condition 7, we know from the proof of Proposition 1.2, that $\cup_{j,k} \ell_j b_1 \ell'_k b_2 L$ is a decomposition of Lb_1b_2L into disjoint right L cosets.

We can write $f_{b_1} = \sum_j \psi(\ell_j)\delta_{\ell_j} * \delta_{b_1} * f_1$, and similarly $f_{b_2} = \sum_k \psi(\ell'_k)\delta_{\ell'_k} * \delta_{b_2} * f_1$. Here f_1 is the identity element of $\mathcal{H}(\mathrm{SL}_2//L,\tilde{\psi})$. In analogy with the argument of Proposition 1.2, we can write $f_{b_1} * f_{b_2} = \sum_{j,k} \psi(\ell_j\ell'_k)\delta_{\ell_j} * \delta_{b_1} * \delta_{\ell'_k} * \delta_{b_2} * f_1$. As noted, thanks to condition 7 we know that the various terms in this sum are supported on disjoint right L cosets, so the only term which contributes to $f_{b_1} * f_{b_2}(b_1b_2)$ is the term $\delta_{b_1} * \delta_{b_2} * f_1$. Hence $f_{b_1} * f_{b_2}(b_1b_2) = f_{b_1b_2}(b_1b_2)$, so the two functions must be equal.

Remark 5.2. Set

$$U^+ = \left\{ \begin{bmatrix} 1 & x \\ 0 & 1 \end{bmatrix} : x \in k \right\} \quad \text{and} \quad U^- = \left\{ \begin{bmatrix} 1 & 0 \\ y & 1 \end{bmatrix} : y \in k \right\}.$$

These are the *root subgroups* relative to the torus A. We also need integral versions of them:

$$U_r^+ = \left\{ \begin{bmatrix} 1 & x \\ 0 & 1 \end{bmatrix} : x \in \mathcal{P}^r \right\} \quad \text{and} \quad U_s^- = \left\{ \begin{bmatrix} 1 & 0 \\ y & 1 \end{bmatrix} : y \in \mathcal{P}^s \right\}.$$

Thus, $U_r^+ = \exp(\mathfrak{g}_{\alpha,r})$, and $U_s^- = \exp(\mathfrak{g}_{-\alpha,s})$.

If $b = \begin{bmatrix} \pi^\ell w & 0 \\ 0 & \pi_{-\ell} w^{-1} \end{bmatrix}$, with w a unit, and $\ell < 0$ (so that b is in A^+), then we can write

$$L_{k,r,s} b L_{k,r,s} = LbL = \bigcup_{y \in \mathcal{P}^s / \mathcal{P}^{s-2\ell}} \begin{bmatrix} 1 & 0 \\ y & 1 \end{bmatrix} bL.$$

Set $\begin{bmatrix} 1 & x \\ 0 & 1 \end{bmatrix} = u_x$ and $\begin{bmatrix} 1 & 0 \\ y & 1 \end{bmatrix} = \bar{u}_y$. Then given b_1 and b_2, with notation parallel to that for b, we see that

$$
\begin{aligned}
Lb_1 L b_2 L &= \bigcup_{y \in \mathcal{P}^s / \mathcal{P}^{s-2\ell_1}} \bigcup_{z \in \mathcal{P}^s / \mathcal{P}^{s-2\ell_2}} \bar{u}_y b_1 \bar{u}_z b_2 L \\
&= \bigcup_{y \in \mathcal{P}^s / \mathcal{P}^{s-2\ell_1}} \bigcup_{z \in \mathcal{P}^s / \mathcal{P}^{s-2\ell_2}} \bar{u}_y \bar{u}_{\pi^{2\ell_1} z} b_1 b_2 L \\
&= \bigcup_{y \in \mathcal{P}^s / \mathcal{P}^{s-2\ell_1}} \bigcup_{z \in \mathcal{P}^s / \mathcal{P}^{s-2\ell_2}} \begin{bmatrix} 1 & 0 \\ y + \pi^{-2\ell_1} z & 1 \end{bmatrix} b_1 b_2 L \\
&= \bigcup_{\tilde{y} \in \mathcal{P}^s / \mathcal{P}^{s-2(\ell_1+\ell_2)}} \bar{u}_{\tilde{y}} b_1 b_2 L.
\end{aligned}
$$

Thus, the decomposition of the product double coset into right cosets can be done very explicitly and cleanly in this situation.

We continue the proof of Proposition 5.2. Observe that, as a semigroup, A is generated by A^+ and by the element $b_1 = \begin{bmatrix} \pi & 0 \\ 0 & \pi^{-1} \end{bmatrix}$. Thus to finish showing that τ is a homomorphism, it is enough to show that $f_{b_1} * f_{b_1^{-1}} = \mu(Lb_1 L)f_1$, or in other words, that $\frac{f_{b_1^{-1}}}{\mu(Lb_1 L)^{1/2}} = \left(\frac{f_{b_1}}{\mu(Lb_1 L)^{1/2}} \right)^{-1}$. As in the remark above, we can write

$$f_{b_1^{-1}} = \sum_{y \in \mathcal{P}^s / \mathcal{P}^{s+2}} \delta_{\bar{u}_y} * \delta_{b_1^{-1}} * f_1 \quad \text{and} \quad f_{b_1} = \sum_{y \in \mathcal{P}^s / \mathcal{P}^{s+2}} f_1 * \delta_{b_1} * \delta_{\bar{u}_y}.$$

Thus, the convolution of the two is

$$f_{b_1} * f_{b_1^{-1}} = \sum_{y,z \in \mathcal{P}^s/\mathcal{P}^{s+2}} f_1 * \delta_{b_1} * \delta_{\overline{u}_y} * \delta_{\overline{u}_z} * \delta_{b_1^{-1}} * f_1$$

$$= \sum_{y,z \in \mathcal{P}^s/\mathcal{P}^{s+2}} f_1 * \delta_{b_1} * \delta_{\overline{u}_{y+z}} * \delta_{b_1^{-1}} * f_1$$

$$=^{\#} (\mathcal{P}^s/\mathcal{P}^{s+2}) \sum_{\tilde{y} \in \mathcal{P}^s/\mathcal{P}^{s+2}} f_1 * \delta_{b_1} * \delta_{u_{\tilde{y}}} * \delta_{b_1^{-1}} * f_1$$

$$= \#(\mathcal{P}^{s-2}/\mathcal{P}^s) \sum_{\tilde{y} \in \mathcal{P}^s/\mathcal{P}^{s+2}} f_1 * \delta_{u_{\tilde{y}}} * f_1.$$

Now the support criterion tells us that in this sum, only the term corresponding to $y = 0$ survives. Thanks to our normalization of Haar measure, this term is just f_1. Since also $\#(\mathcal{P}^{s-2}/\mathcal{P}^s) = \mu(Lb_1 L)$, this gives us the relation we wanted.

Finally, let us demonstrate the support criterion, statement (i) of Proposition 5.2. This involves a fairly simple calculation plus an appeal to Proposition 5.1. The basis of the calculation is contained in the following lemma.

Lemma 5.1. $Ad(L)(ah + (\Lambda_{k,r,s}^\perp \cap \mathfrak{t})) = ah + \Lambda_{k,r,s}^\perp$.

Proof. The proof is an easy exercise using Hensel's Lemma.

Suppose that $g \in \mathrm{Supp}(\mathcal{H}(\mathrm{SL}_2//L, \tilde{\psi}))$. Then Proposition 5.2 says that we can find an element $z \in Ad(g)(ah + \Lambda_{k,r,s}^\perp) \cap (ah + \Lambda_{k,r,s}^\perp)$. ¿From Lemma 5.1, we can write $z = Ad(h_1)(y)$, with $y \in ah + (\Lambda_{k,r,s}^\perp \cap \mathfrak{t})$. Similarly, we can also write $z = Ad(g)(Ad(hl_2))(y')$ for some ℓ_2 in L and y' in $ah + (\Lambda_{k,r,s}^\perp \cap \mathfrak{t})$. Comparing these two expressions for z, we see that $Ad(h_1)(y) = Ad(g)Ad(h_2)(y')$, or $Ad(h_1^{-1}gh_2)(y') = y$.

Notice that, because of the requirement that $\ell > k$, the set $ah + \Lambda_{k,r,s}^\perp$ does not contain the zero element of $\mathfrak{t}$. This also implies that, since $ah + \Lambda_{k,r,s}^\perp$ is "convex", in the sense that it contains the average of any two of its elements, if it contains an element y, it does not contain $-y$. But the only elements of $\mathfrak{t}$ which are conjugate to some y in $\mathfrak{t}$ are y itself and $-y$. Thus, we conclude that $g' = h_1^{-1}gh_2$ centralizes y. Since $\mathfrak{t}$ consists simply of multiples of y, this means that g' centralizes $\mathfrak{t}$, and hence belongs to A, or in other words $g \in LAL$, as desired. This concludes the proof of Proposition 5.2.

Remark 5.3.
The key point that makes Proposition 5.2 work is that we chose r and s to make $r + s = \ell$, which is the largest possible value such that the Kirillov form vanishes on $\Lambda_{k,r,s}$, so that we can use $\psi_{ah} \circ \log$ to define a character on $L = exp(\Lambda_{k,r,s})$. The fact that $r + s = \ell$ is what makes Lemma 5.3.2 true, and this is in turn is what makes part $i)$ of Proposition 5.2 work. Further, as we saw in the proof, the second part of Proposition 5.2, which is what we really

want to know, depends crucially on the first part. We note again that there
are many possible pairs r, s that will work; the essential condition is on their
sum.

Let $A_o = A^+ \cap A^-$ be the maximal compact subgroup of A. It is easy to see
that A_o normalizes L, so that the product $A_o L$ will be a group. Further, the
character ψ_{ah} of L, extends to $A_o L$. The extensions are naturally identified
with the extensions from $A \cap L$ to A_o of $\psi|_{A \cap L}$.

Denote one such extension by Ψ. Then easy arguments show that $\mathcal{H}(A//A_o, \Psi)$
$\simeq C_c(A/A_o) \simeq C_c(k^\times/\mathcal{O}) \simeq C_c(\mathbb{Z})$. This may be regarded as the simplest
possible example of an affine Hecke algebra. Furthermore, an easy corollary
of Proposition 5.2 is that $\mathcal{H}(\mathrm{SL}_2//A_o L, \Psi) \simeq \mathcal{H}(A//A_o, \Psi)$. Thus, Proposition 5.2 is a prototype for more general Hecke algebra isomorphisms.

We note that the spherical function algebra $\mathcal{H}(\mathrm{SL}_2//A_o L, \Psi)$ controls the
non-spherical principal series of SL_2 corresponding to all characters of A
which restrict to Ψ on A_o. Also, one can show that $\mathcal{H}(\mathrm{SL}_2//L, \tilde{\psi})$ is isomorphic to the direct sum of the algebras $\mathcal{H}(\mathrm{SL}_2//A_o L, \Psi)$ where Ψ runs over all
possible extensions of ψ_{ah} from $A \cap L$ to A_o. Thus, $\mathcal{H}(\mathrm{SL}_2//L, \tilde{\psi})$ controls
the union of a collection of principal series representations of SL_2.

For completeness in the case of SL_2, one should also analyze spherical function
algebras $\mathcal{H}(\mathrm{SL}_2//J, \phi)$, where ϕ is a character of $J/J_1 \simeq A_o/A_1 \simeq \overline{k}^\times$. Here
one can show that if $\phi \neq \phi^{-1}$, then $\mathcal{H}(\mathrm{SL}_2//J, \phi) \simeq C_c(\mathbb{Z})$. In the case of
the sign character $\mathrm{sgn} = \mathrm{sgn}^{-1}$, one has $\mathcal{H}(\mathrm{SL}_2//J, \mathrm{sgn}) \simeq C_c(\tilde{W}_{\mathrm{SL}_2})$ is the
group algebra of the affine Weyl group of SL_2, which is the infinite dihedral
group.

5.4 General minimal principal series

The example described in detail in Section 5.3 is not so different from the
general case, at least not when p is sufficiently large.

Let $\mathfrak{g}$ be a split semisimple Lie algebra, and let $\mathfrak{t} \subset \mathfrak{g}$ be a split Cartan
subalgebra. Write

$$\mathfrak{g} = \mathfrak{t} \oplus \sum_\alpha \mathfrak{g}_\alpha$$

for the root space decomposition of $\mathfrak{g}$ with respect to $\mathfrak{t}$. The $\mathfrak{g}_\alpha$ are the root
spaces - the eigenspaces for the action $\mathrm{ad}\mathfrak{t}$ of $\mathfrak{t}$ on $\mathfrak{g}$ by commutator.

Let $\mathfrak{t}_o \subset \mathfrak{t}$ be the lattice in $\mathfrak{t}$ consisting of elements x such that the
eigenvalues of $\mathrm{ad}(x)$ are integers. We can find a Lie lattice $\Lambda \subset \mathfrak{g}$, containing
$\mathfrak{t}_o$. Under the assumption that the residual characteristic p is large, we will
have

$$\Lambda = \mathfrak{t}_o \oplus \sum_\alpha \Lambda \cap \mathfrak{g}_\alpha$$

for any lattice containing $\mathfrak{t}_o$ and invariant under $\mathrm{ad}\mathfrak{t}_o$.

Recall that the Killing form $\kappa(\ ,\ )$ on $\mathfrak{g}$ is defined by the formula

$$\kappa(x,y) = tr(\mathrm{ad}x, \mathrm{ad}y).$$

An element x of a Lie lattice such as Λ will of course satisfy $\mathrm{ad}x(\Lambda) \subset \Lambda$. It follows that $\kappa(x,y)$ must also be integral. Thus, letting Λ^* denote the lattice dual to Λ by the Killing form, we will have $\Lambda \subset \Lambda^*$. Thus, if we have a Lie lattice in $\mathfrak{g}$ such that $\Lambda^* = \Lambda$, it must be a maximal Lie lattice. Chevalley's description of the generators and relations for a simple Lie algebra in terms of roots shows that when p is large (actually, it does not have to be very large for this), self-dual Lie lattices do exist in any semi-simple Lie algebra $\mathfrak{g}$. For $\mathfrak{g} = \mathfrak{sl}_n$, the stabilizer of any lattice is a self-dual Lie lattice, and for $\mathfrak{g} = \mathfrak{sp}_{2n}$, the stabilizer of any *self-dual* lattice is a self-dual Lie lattice.

For any fixed root α, set

$$\mathfrak{t}_\alpha = [\mathfrak{g}_\alpha, \mathfrak{g}_{-\alpha}].$$

This is a line in $\mathfrak{t}$. The span

$$\mathfrak{s}_\alpha = \mathfrak{t}_\alpha \oplus \mathfrak{g}_\alpha \oplus \mathfrak{g}_{-\alpha}$$

is a Lie algebra, isomorphic with $\mathfrak{sl}_2$. We have a decomposition $\mathfrak{g} = \mathfrak{s}_\alpha \oplus \mathfrak{s}_\alpha^\perp$, where the $\perp$ is with respect to κ.

Suppose that $\Lambda = \mathfrak{t}_o + \sum_\alpha \Lambda \cup \mathfrak{g}_\alpha$ is a self-dual Lie lattice in $\mathfrak{g}$. Then the intersection $\mathfrak{s}_\alpha \cap \Lambda$ is a self-dual Lie lattice in $\mathfrak{s}_\alpha$, and $\Lambda = \mathfrak{s}_\alpha \cap \Lambda \oplus \mathfrak{s}_\alpha^\perp \cap \Lambda$. Set $\mathfrak{g}_{\alpha,o} = \mathfrak{g}_\alpha \cap \Lambda$ and $\mathfrak{t}_{\alpha,o} = \mathfrak{t}_\alpha \cap \Lambda$. The fact that Λ is self-dual implies that $[\mathfrak{g}_{\alpha,o}, \mathfrak{g}_{\beta,o}] = \mathfrak{g}_{\alpha+\beta,o}$ whenever $\alpha+\beta$ is a root. Similarly, $[\mathfrak{g}_{\alpha,o}, \mathfrak{g}_{-\alpha,o}] = \mathfrak{t}_{\alpha,o}$. As in Section 5.3, for an integer ℓ, set $\Lambda_\ell = \pi^\ell \Lambda$, and $\mathfrak{t}_\ell = \pi^\ell \mathfrak{t}_o$, and $\mathfrak{g}_{\alpha,\ell} = \pi^\ell \mathfrak{g}_{\alpha,o}$ and $\mathfrak{t}_{\alpha,\ell} = \pi^\ell \mathfrak{t}_{\alpha,o}$.

Let ψ be a character of $\mathfrak{t}_1$. We can represent ψ in the form $\psi(x) = \chi(\kappa(x, y_\psi))$, where χ is a character of (the additive group of) k, such that $\mathcal{O} \subset \ker \chi$, but $\pi^{-1}\mathcal{O} \not\subset \ker \chi$. By the *depth* of ψ, denoted d_ψ, we mean the smallest ℓ such that $\mathfrak{t}_\ell \subset \ker \psi$. This is also equal to the smallest ℓ such that $y_\psi \in \mathfrak{t}_{-\ell}$. Similarly define $d_{\psi,\alpha}$ to be the smallest ℓ such that $\mathfrak{t}_{\alpha,\ell} \subset \ker \psi$. Notice that $d_\psi = \max_\alpha d_{\psi,\alpha}$ The notion of depth extend in a straightforward way from characters of $\mathfrak{t}_1$ to characters of $\mathfrak{t}_k$.

Lemma 5.2. $d_{\psi,\alpha+\beta} \leq max\{d_{\psi,\alpha}, d_{\psi,\beta}\}$

The proof uses that $\mathfrak{t}_{\alpha+\beta,\ell} \subset \mathfrak{t}_{\alpha,\ell} + \mathfrak{t}_{\beta,\ell}$, which can be shown by considering rank 2 algebras. In fact for the classical cases, only $\mathfrak{sl}_3$ and $\mathfrak{sp}_4$ need to be considered.

Corollary 5.1. *For any fixed d_0, $\{\beta \mid d_{\psi,\beta} \leq d_0\}$ span a Levi component M_{ψ,d_0}.*

The Levi component M_{ψ, d_0} in the Corollary consists of all directions where the character ψ is trivial. In general the roots where a particular character has depth less than or equal to a specified level, say j, forms a Levi subalgebra and every Levi subalgebra can be described in this way. If such an $M_{\psi, j} = \mathfrak{t}$ then we say ψ is nondegenerate at level j.

Choose an ordering for the roots. Set

$$\mathfrak{J} = \mathfrak{t}_1 \oplus \sum_{\alpha > 0} \mathfrak{g}_{\alpha, o} \oplus \sum_{\alpha < 0} \mathfrak{g}_{\alpha, 1}.$$

Then $\mathfrak{J}$ is the Lie lattice corresponding to an Iwahori subgroup of G. Given any character ψ of $\mathfrak{t}_1$, we may extend ψ to a character of $\mathfrak{J}$ by declaring ψ to be trivial on any root space: $\psi(x) = 1$, for all x in $\mathfrak{g}_\alpha \cap \mathfrak{J}$. Define a function on $\mathfrak{J} \times \mathfrak{J}$ by the formula

$$B_\psi(x, y) = \psi([x, y])$$

for x and y in $\mathfrak{J}$. This is the Kirillov bicharacter associated to ψ.

Analogous to the $\mathfrak{sl}_2$ case, consider the Kirillov bicharacter associated to ψ and construct a Lie lattice Λ_ψ maximal with respect to the condition $B_\psi|_{\Lambda_\psi} = 1$. ($B_\psi$ breaks into a sum of bicharacters on copies of $\mathfrak{sl}_2$.) For each α such that $d_{\psi, \alpha} > j$, choose $a_\alpha, a_{-\alpha} \geq 0$ so that $a_\alpha + a_{-\alpha} = d_{\psi, \alpha}$. If $d_{\psi, \alpha} = j$, set $a_\alpha = a_{-\alpha} = j$. We also require that $\Lambda_\psi = \mathfrak{t}_k \oplus \sum_\alpha \mathfrak{g}_{\alpha, a_\alpha}$ be a Lie lattice ($a_{\alpha + \beta} \leq a_\alpha + a_\beta$). There do exist such possibilities for a_α. For example, take $a_\alpha = 0$ for $\alpha > 0$ and $a_{-\alpha} = d_{\psi, \alpha}$, or take $a_\alpha = \left\lfloor \dfrac{d_{\psi, \alpha}}{2} \right\rfloor$ for $\alpha > 0$ and $a_{-\alpha} = d_{\psi, \alpha} - a_\alpha = \left\lfloor \dfrac{d_{\psi, \alpha} + 1}{2} \right\rfloor$.

Extend ψ to a character of Λ_ψ by declaring ψ to be trivial on $\mathfrak{g}_{\alpha, a_\alpha}$. Then $\tilde{\psi} = \psi \circ \log$ is a character of $L_\psi = \exp \Lambda_\psi$. Let $M_{\psi, 1} = M_\psi$ be as in Corollary 5.1.

Lemma 5.3. *If $g \in \mathrm{Supp}(\mathcal{H}(G//L_\psi, \tilde{\psi}))$, then $g \in L_\psi M_\psi L_\psi$.*

Let $m \in M_\psi$ and define $f_m(k_1 m k_2) = \tilde{\psi}(k_1 k_2)$ for $k_1, k_2 \in L_\psi$, and f_m to have value 0 off of $L_\psi m L_\psi$. Then set $\tilde{f}_m = f_m|_{M_\psi}$.

Theorem 5.1. *With $\tilde{L}_\psi = L_\psi \cap M_\psi$, the map*

$$\tau : f_m \to \left(\frac{\mu_G(L_\psi)\mu_G(L_\psi m L_\psi)}{\mu_{M_\psi}(\tilde{L}_\psi)\mu_{M_\psi}(\tilde{L}_\psi m \tilde{L}_\psi)} \right)^{1/2} \tilde{f}_m$$

is an isomorphism of algebras and is, up to a constant multiple, an L^2 isometry.

The proof is nearly the same as for $G = \mathrm{SL}_2$ above.

6 Consequences

The discussion in Section 5.4 implies the minimal principal series can be analyzed by reduction to $\mathcal{H}(G//J, \phi)$ for $\phi \in \hat{J}$. This is done by eliminating the co-root directions where the character has positive depth on the torus and focusing on where the depth is zero. Such algebras have been analyzed by A. Roche and shown to be affine Hecke algebras. In fact these ideas can be extended much further.

Recall that $\Lambda_o = \mathfrak{t}_o \oplus \sum_\alpha \mathfrak{g}_{\alpha,o}$ is a lattice corresponding to a good maximal parahoric subgroup. Also recall that

$$\mathfrak{J} = \mathfrak{t}_o \oplus \sum_{\alpha>0} \mathfrak{g}_{\alpha,o} \oplus \mathfrak{g}_{-\alpha,1}$$

plays the role of the Iwahori lattice. Choose a Lie lattice $\check{\Lambda}$ such that $\mathfrak{J} \subset \check{\Lambda} \subset \Lambda_o$ to play the role of the parahoric lattice. Let Γ be a finite subgroup of $\mathrm{Aut}G$ with $|\Gamma| < p$ (although for many of the results, any cardinality relatively prime to p is sufficient). Then $\mathfrak{g} = \mathfrak{g}^\Gamma \oplus \mathfrak{h}$ and we define the projection map $\mathrm{pr}_\Gamma : \mathfrak{g} \to \mathfrak{g}^\Gamma$ by

$$\mathrm{pr}_\Gamma(x) = \frac{1}{|\Gamma|} \sum_{\gamma \in \Gamma} \gamma(x).$$

Notice that $[\mathfrak{g}^\Gamma, \mathfrak{h}] \subset \mathfrak{h}$. Suppose that $\check{\Lambda}$ and $\mathfrak{t}$ are stable under the action of Γ. Then

$$\gamma(\mathfrak{g}_\alpha) = \mathfrak{g}_{\gamma(\alpha)} \quad \text{and} \quad \gamma(\mathfrak{g}_{\alpha,\epsilon(\alpha)}) = \mathfrak{g}_{\gamma(\alpha),\epsilon(\gamma(\alpha))},$$

where $\epsilon(\alpha)$ is 0 or 1 as appropriate.

Now take $\psi \in (\hat{\mathfrak{t}}_k)^\Gamma \simeq \widehat{\mathfrak{t}_k^\Gamma} \hookrightarrow \widehat{\check{\Lambda}_k}$ (the isomorphism holds since the order of Γ is relatively prime to p). Then the depths will be constant on the orbits of γ, i.e. $d_{\psi,\alpha} = d_{\psi,\gamma(\alpha)}$. Construct Λ_ψ to be Γ invariant. If $-\alpha$ and α are in the same orbit under Γ then to do this we must adjust the α and $-\alpha$ root spaces simultaneously which may not be consistent with the requirement that $a_\alpha + a_{-\alpha} = d_{\psi,\alpha}$, so this may not quite be possible. The best possible may only be $a_\alpha + a_{-\alpha} = d_{\psi,\alpha} \pm 1$. If so, define $\Lambda_\psi^\sharp$ using $d_{\psi,\alpha} - 1$ and $\Lambda_\psi^\flat$ using $d_{\psi,\alpha} + 1$ (so $\Lambda_\psi^\flat \subset \Lambda_\psi^\sharp$) as well as $L_\psi^\sharp = \exp \Lambda_\psi^\sharp$ and $L_\psi^\flat = \exp \Lambda_\psi^\flat$. Then

1. $\tilde{\psi} = \psi \circ \log$ is a character of $L_\psi^\flat$.

2. $\tilde{\psi}$ is $\mathrm{Ad}L_\psi^\sharp$ invariant.

3. There exists a unique character $\rho_\psi \in \widehat{L_\psi^\sharp}$ such that $\rho_\psi|_{L_\psi^\flat} = m\tilde{\psi}$ where

$$m = \dim \rho_\psi = \left| \frac{L_\psi^\sharp}{L_\psi^\flat} \right|^{1/2}.$$

Modulo the kernel of $\tilde{\psi}$ we have the Heisenberg group and we take representations of it. We get a similar result to that before for fixed points, namely

$(L^b_\psi)^\Gamma = L^b_\psi \cap G^\Gamma = \exp \Lambda^{b\Gamma}_\psi$ and $L^b_\psi = (L^b_\psi)^\Gamma \oplus (L^b_\psi \cap \mathfrak{h})$ with a similar result for the case of $L^\sharp_\psi$.

Proposition 6.1. $\mathrm{Supp}(\mathcal{H}(G^\Gamma/(L^\sharp_\psi)^\Gamma, \rho^\Gamma_\psi)) = (L^\sharp_\psi)^\Gamma M^\Gamma_\psi (L^\sharp_\psi)^\Gamma$

It is an open question whether there is an analogous Hecke algebra isomorphism associated with this support result. If so, then this may complete 80-90% of the theory. This was done in some cases for $G = GLn$ Howe and Moy, but has not been pushed further.

Example 6.1. Let G be a reductive group over k. Let $\tilde{k}$ be a Galois extension over k which splits some or possibly all of the tori in G_k. Let $\Gamma = \mathrm{Gal}(\tilde{k}, k)$ which acts on $G_{\tilde{k}}$ and the Lie algebra $\mathfrak{g}_{\tilde{k}} = \mathfrak{g}_k \otimes \tilde{k}$.

Lemma 6.1. *If* $\mathfrak{t} \subset \mathfrak{g}_k$ *and* $\mathfrak{t} \otimes \tilde{k}$ *is split, then we can find a* Γ-*stable Lie lattice* $\tilde{\Lambda}$ *for* $\mathfrak{t} \otimes \tilde{k} \subset \mathfrak{g}_{\tilde{k}}$.

This has been verified when G is classical. Then M^Γ is the centralizer of a torus and can be compact. In the extreme case where Γ is nondegenerate, M^Γ is a torus and again might be compact. If M^Γ is compact then the corresponding spherical function algebra is finite dimensional, and the representation induced from such ψ will be a finite sum of irreducible supercuspidal representations. This case of the construction yields all known supercuspidal representations. Supercuspidal representations which correspond to characters of compact tori were constructed by Morris The above construction extends Morris's work, and explains the origin of other supercuspidal representations that do not correspond to characters of compact tori.

Example 6.2. If $G = \mathrm{Sp}_{2n}$ then M^Γ is a product of unitary groups, copies of GL_k, and possibly one copy of Sp_{2m}. This yields supercuspidal representations corresponding to characters of a non-maximal torus whose centralizer is a product of compact unitary groups.

At the moment, this kind of construction accounts for all known examples of supercuspidal representations, at least in the regime of large residual characteristic. For the full story, what we have done here must be further refined, by extending ρ^Γ_ψ to a cuspidal representation on a parahoric subgroup in M^Γ. For Sp_{2n}, this more refined construction yields supercuspidal representations when the centralizer of a torus is contains no GL_k factors.

One can ask whether the Hecke algebras that arise in this way are all generalized affine Hecke algebras. The answer is yes in many cases. For the case of GL_n this was settled by Howe-Moy and Bushnell-Kutzko (with no restriction on characters). For algebras associated to level zero, this was done by Morris. A. Roche has obtained results for principal series, and many new examples have been provided by J. Kim.

These notes represent an adaptation and summary or synthesis of a range of papers. Rather than give precise references, we list below for each section a selection of papers which serve either as background or development of the content of the notes.

It is a pleasure to thank Cathy Kriloff for shouldering the main burden in preparing the text.

References

2. Structure of p-adic GL(V) and Sp(V)

[BT]   F. Bruhat and J. Tits, Groupes réductifs sur un corps local I. *Inst. Hautes Études Sci. Publ. Math.* **42** (1972), 1–251.

3. Structure of the Iwahori Hecke Algebra

[BZ]   J. Bernstein and A. Zelevinsky, Representations of the group GL(n, F) where F is a non-archimedean local field. *Russian Math. Surveys* **31** (1976), 1–68.

[Bo]   A. Borel, Admissible representations of a semisimple group over a local field with vectors fixed under an Iwahori subgroup. *Invent. Math.* **35** (1976), 233–259.

[Ca]   W. Casselman, The unramified principal series of p-adic groups. I. The spherical function. *Compositio Math.* **40** (1980), 387–406.

[Ch]   I. Cherednik, A new interpretation of Gelfand-Tsetlin bases. *Duke Math. J.* **54** (1987), 563–577.

[H4]   R. Howe, Hecke algebras and p-adic GL$_n$. *Contemp. Math.* **177** (1994), 65–100.

[KL]   D. Kazhdan and G. Lusztig, Proof of the Deligne-Langlands conjecture for Hecke Algebras. *Invent. Math.* **87** (1987), 153–215.

[L1]   G. Lusztig, Some examples of square integrable representations of semisimple p-adic groups. *Trans. Amer. Math. Soc.* **277** (1983), 623–653.

[L2]   —, Affine Hecke algebras and their graded version. *J. Amer. Math. Soc.* **2** (1989), 599–635.

[Ro]   J. Rogawski, On modules over the Hecke algebra of a p-adic group. *Invent. Math.* **79** (1985), 443–465.

4. Results on Representations of G

[BM1] D. Barbasch and A. Moy, A unitarity criterion for p-adic groups. *Invent. Math.* **98** (1989), 19–37.

[BM2] —, Reduction to real infinitesimal character in affine Hecke algebras. *J. Amer. Math. Soc.* **6** (1993), no. 3, 611–635.

[BM3] —, Whittaker Models with Iwahori Fixed Vector. *Contemporary Math.* **177** (1994), ams

[BM4] —, Unitary Spherical Spectrum for p-adic classical groups. *Acta Appl. Math.* **44**, Kluwer Academic Press, 1996.

[C]   P. Cartier, Representations of p-adic groups: a survey. *Automorphic Forms, Representations and L-functions*, A. Borel and W. Casselman (eds.), Proc. Symp. Pure Math. XXXIII, Parts 1 & 2, American Mathematical Society, Providence, RI, 1979, 111–155.

[Ca] W. Casselman, The unramified principal series of p-adic groups I. The spherical function. *Compositio Math.* **40** (1980), 387–406.

5. Spherical Function Algebras

[BK1] C. Bushnell and P. Kutzko, *The Admissible Dual of* GL(N) *via Compact Open Subgroups*, Ann. of Math. Stud., vol. 129, Princeton Univ. Press, 1993.

[H1] R. Howe, Kirillov theory for compact p-adic groups. *Pacific J. Math.* **73** (1977), 365–381.

[H4] —, Hecke algebras and p-adic GL_n. *Contemp. Math.* **177** (1994), 65–100.

[R]   A. Roche, Types and Hecke Algebras for principal series representations of split reductive p-adic groups. *Ann. Sci. École Norm. Sup.* **31** (1998), 361–413.

6. Consequences

[A]   J. Adler, Refined anisotropic K-types and supercuspidal representations. *Pacific J. Math.* **185** (1998), 1–32.

[As] C. Asmuth, Weil representations of symplectic p-adic groups. *Amer. J. Math.* **101** (1979), 885–908.

[BK1] C. Bushnell and P. Kutzko, *The Admissible Dual of* GL(N) *via Compact Open Subgroups*, Ann. of Math. Stud., vol. 129, Princeton Univ. Press, 1993.

[BK2] —, Smooth representations of reductive p-adic groups: Structure theory via types. *Proc. London Math. Soc.* (3) **77** (1998), no. 3, 582–634.

[H2] R. Howe, Some qualitative results on the representation theory of Gl_n over a p-adic field. *Pacific J. Math.* **73** (1977), 479–538.

[H3] —, Tamely ramified supercuspidal representations of Gl_n. *Pacific J. Math.* **73** (1977), 437–560.

[H4] —, Hecke algebras and p-adic GL_n. *Contemp. Math.* **177** (1994), 65–100.

[HM1] R. Howe (with A. Moy), *Harish-Chandra Homomorphisms for p-adic Groups*, CBMS Regional Conf. Ser. in Math., vol. 59, Amer. Math. Soc., Providence, RI, 1985.

[HM2] R. Howe and A. Moy, Hecke algebra isomorphisms for $GL(n)$ over a p-adic field. *J. Alg.* **131** (1990), 388–424.

[K]   J.-L. Kim, Hecke algebras of classical groups over p-adic fields and supercuspidal representations. *Amer. J. Math.* **121** (1999), no. 5, 967–1029.

[L3] G. Lusztig, Classification of unipotent representations of simple p-adic groups. *Internat. Math. Res. Notices* **11** (1995), 517–589.

[M1] L. Morris, Some tamely ramified supercuspidal representations of symplectic groups. *Proc. London Math. Soc.* **63** (1991), 519–551.

[M2] —, Tamely ramified supercuspidal representations of classical groups II. Representation theory. *Ann. Sci. Éc. Norm. Sup.* **25** (1992), 233–274.

[M3] —, Tamely ramified intertwining algebras. *Invent. Math.* **114** (1994), 1–54.

[MP1] A. Moy and G. Prasad, Unrefined minimal K-types for p-adic groups. *Invent. Math.* **116** (1994), 393–408.

[MP2] —, Jacquet Functors and unramified minimal K-types. *Comment. Math. Helv.* **71** (1996), 98–121.

[Wa] J.-L. Waldspurger, Algébras de Hecke et induites de représentations cuspidales, pour GL(N). *J. Reine Angew. Math.* **370** (1986), 127–191.

[Y] J.-K. Yu, Tame construction of supercuspidal representations. *preprint*, 1998.

Notes on affine Hecke algebras

George Lusztig

Department of Mathematics, M.I.T., Cambridge, MA 02139
gyuri@math.mit.edu
Supported in part by the National Science Foundation

Introduction

Affine Hecke algebras play an important role in the study of representations of reductive groups over a p-adic field. They also encode critical information about the characters of modular representations of semisimple algebraic groups in positive characteristic and of quantum groups at roots of 1. One approach to affine Hecke algebras is via equivariant K-theory. In this approach, the parameter of the algebra is interpreted as the standard generator of the representation ring of $\mathbb{C}^*$. In these lectures, the K-theoretic approach to the affine Hecke algebras is emphasized. We will use this approach to give an exposition of the classification of the simple modules of the affine Hecke algebra. We will also give a survey of the results of [L6] on bases in equivariant K-theory (subregular case) and will extend these results to the not necessarily simply laced case.

Contents

1 The affine Hecke algebra

1.1

Let X, Y be two finitely generated free abelian groups with a given perfect pairing $\langle,\rangle : X \times Y \to \mathbb{Z}$. Assume that $\{\alpha_i | i \in I\} \subset X$ (resp. $\{\check{\alpha}_i | i \in I\} \subset Y$) are the simple roots (resp. simple coroots) of a root datum of finite type in X, Y. For $i \in I$ let $s_i : X \to X$ be the reflection $s_i(x) = x - \langle x, \check{\alpha}_i \rangle \alpha_i$ and let W be the subgroup of $GL(X)$ generated by $s_i, i \in I$ (the Weyl group, a Coxeter group).

Let $\mathcal{A} = \mathbb{Z}[v, v^{-1}]$ where v is an indeterminate. Let $\mathcal{A}[X]$ be the group algebra of X with coefficients in $\mathcal{A}$. The basis element of $\mathcal{A}[X]$ corresponding to x is denoted by $[x]$. Let $\mathcal{H}$ be the $\mathcal{A}$-algebra with 1 defined by the generators

$\tilde{T}_i (i \in I)$ and $\theta_x (x \in X)$

and the relations

(a) $(\tilde{T}_i + v^{-1})(\tilde{T}_i - v) = 0$ for any $i \in I$;

(b) $\tilde{T}_i \tilde{T}_j \tilde{T}_i \cdots = \tilde{T}_j \tilde{T}_i \tilde{T}_j \ldots$

(μ factors in both products), for any $i \neq j$ in I with $s_i s_j$ of order μ in W;

(c) $\theta_x \theta_{x'} = \theta_{x+x'}$ for any $x, x' \in X$;

(d) $\theta_0 = 1$;

(e) $\theta_x \tilde{T}_i - \tilde{T}_i \theta_{s_i(x)} = (v - v^{-1}) \theta_{\frac{[x] - [s_i(x)]}{1 - [-\alpha_i]}}$

for any $i \in I$, $x \in X$.

The fraction in (e) is apriori an element of the quotient field of $\mathcal{A}[X]$ but it actually belongs to $\mathcal{A}[X]$; for $p = \sum_{x \in X} c_x[x] \in \mathcal{A}[X]$ with $c_x \in \mathcal{A}$ we write $\theta_p = \sum_{x \in X} c_x \theta_x \in \mathcal{H}$.

1.2

More generally, assume that we are given functions

$$\lambda : I \to \mathbb{N}, \quad \lambda^* : \{i \in I | \check{\alpha}_i \in 2Y\} \to \mathbb{N}$$

such that $\lambda(i) = \lambda(i')$ whenever $i, i' \in I$ satisfy $\langle \alpha_{i'}, \check{\alpha}_i \rangle = \langle \alpha_i, \check{\alpha}'_i \rangle = -1$.

To λ, λ^* we can attach the $\mathcal{A}$-algebra $\mathcal{H}$ with 1 defined by the generators $\tilde{T}_i (i \in I)$ and $\theta_x (x \in X)$ and the relations 1.1(b)-(d) together with

(a) $(\tilde{T}_i + v^{-\lambda(i)})(\tilde{T}_i - v^{\lambda(i)}) = 0$ for any $i \in I$;

(b) $\theta_x \tilde{T}_i - \tilde{T}_i \theta_{s_i(x)} = (v^{\lambda(i)} - v^{-\lambda(i)}) \theta_{\frac{[x] - [s_i(x)]}{1 - [-\alpha_i]}}$

for any $i \in I$, $x \in X$ with $\check{\alpha}_i \notin 2Y$;

(c) $\theta_x \tilde{T}_i - \tilde{T}_i \theta_{s_i(x)} = ((v^{\lambda(i)} - v^{-\lambda(i)}) + (v^{\lambda^*(i)} - v^{-\lambda^*(i)}) \theta_{-\alpha_i}) \theta_{\frac{[x] - [s_i(x)]}{1 - [-2\alpha_i]}}$

for any $i \in I$, $x \in X$ with $\check{\alpha}_i \in 2Y$. The fractions are interpreted as in 1.1.

In the "special case" where $\lambda(i) = 1$ for all i and $\lambda^*(i) = 1$ for all i such that $\lambda^*(i)$ is defined, we recover the algebra in 1.1.

1.3

$\mathcal{H}$ is called an *affine Hecke algebra*. It is another incarnation of the Hecke algebra of an (extended) affine Weyl group introduced by Iwahori and Matsumoto [IM] in which the parameters attached to the various (affine) simple reflections are allowed to depend on the simple reflection. The fact that the Hecke algebra in the Iwahori-Matsumoto presentation can be described in the form above has been stated by Bernstein (in the "special case" 1.1); the proof was given in [L1] in the "special case" and in [L3] in general. The existence of the two incarnations of $\mathcal{H}$ reflects (in the "special case" 1.1) the fact that an algebraic group over a p-adic field is at the same time an infinite dimensional group over a finite field. A strategy for classifying irreducible representations of a semisimple split p-adic group, say, with non-zero vectors fixed by an Iwahori group, has been to convert the problem to that of the classification of irreducible representations of an affine Hecke algebra in the

Iwahori-Matsumoto presentation (with v specialized to a power of $\sqrt{p}$), then to pass to the presentation in 1.1 which is particularly well suited for the study of representations.

1.4

For $w \in W$ we set
$$\tilde{T}_w = \tilde{T}_{i_1}\tilde{T}_{i_2}\ldots\tilde{T}_{i_n} \in \mathcal{H}$$
where $i_1, i_2, \ldots, i_n$ is any sequence in S such that $w = s_{i_1}s_{i_2}\ldots s_{i_n}$ and such that n is minimum possible; this is known to be independent of the choice of sequence. Recall that n is the length, $l(w)$ of w.

$\mathcal{H}$ is free as an $\mathcal{A}$-module. Indeed, the elements $\tilde{T}_w\theta_x$ (with $w \in W, x \in X$) form an $\mathcal{A}$-basis of $\mathcal{H}$; the elements $\theta_x\tilde{T}_w$ (with $w \in W, x \in X$) form another $\mathcal{A}$-basis of $\mathcal{H}$. (See [[L3] 3.4].)

1.5

Let w_0 be the longest element of W. Let $\chi \mapsto \chi^\blacktriangle$ be the involutive antiautomorphism of the $\mathcal{A}$-algebra $\mathcal{H}$ defined by $\tilde{T}_i \mapsto \tilde{T}_i$ for $i \in I$ and $\tilde{T}_{w_0}^{-1}\theta_{w_0(x)}\tilde{T}_{w_0} \mapsto \theta_{-x}$ for $x \in X$.

2 $\mathcal{H}$ and equivariant K-theory

2.1

The idea that the representations of $\mathcal{H}$ (in the "special case" 1.1) should be geometrically understood in terms of equivariant K-theory, the parameter v being interpreted as the standard generator of the representation ring of $\mathbb{C}^*$, was formulated in [L2], where the principal series representations of $\mathcal{H}$ were treated from this point of view. Subsequently, this idea has been developed in [KL], [GI].

In this section we want to explain the method of [KL] to construct representations of $\mathcal{H}$. We shall need some K-theoretical preparation.

2.2

All algebraic varieties are assumed to be over $\mathbb{C}$. If M is a linear algebraic group we say that Z is an M-variety if M acts on Z and Z can be imbedded as a closed M-stable subvariety of a smooth algebraic variety with an algebraic action of M. For such Z, let $Coh_M(Z)$ be the category of coherent M-equivariant sheaves on X and let $Vec_M(Z)$ be the full subcategory of $Coh_M(Z)$ whose objects are locally free coherent sheaves. (We identify objects of $Vec_M(Z)$ with the corresponding M-equivariant vector bundles on

Z.) Let $K_M(Z)$ be the Grothendieck group of the category $Coh_M(Z)$. This is naturally an R_M-module where $R_M = K_M(\text{point})$ is the Grothendieck ring of finite dimensional representations of M.

Let Z' be another M-variety and let $f : Z \to Z'$ be an M-equivariant morphism. If f is smooth, then the inverse image $f^* : K_M(Z') \to K_M(Z)$ is well defined; if f is proper, the the direct image $f_* : K_M(Z) \to K_M(Z')$ is well defined (it is defined using an alternating sum of higher direct images).

If Z' is a closed M-stable subvariety of Z, let $Coh_M(Z; Z')$ be the full subcategory of $Coh_M(Z)$ whose objects are the objects of $Coh_M(Z)$ whose support is contained in Z'. Let $K_M(Z; Z')$ be the Grothendieck group of the category $Coh_M(Z; Z')$. We have an obvious isomorphism $K_M(Z') \xrightarrow{\sim} K_M(Z; Z')$.

Let $\phi : E_1 \to E_0$ be a morphism in $Vec_M(Z)$ and let Z' be a closed subvariety of Z such that ϕ is an isomorphism over $Z - Z'$. Let $\mathcal{F} \in Coh_M(Z)$. Let $\mathcal{K}_1, \mathcal{K}_0$ be the kernel and cokernel of the morphism $1 \otimes \phi : \mathcal{F} \otimes E_1 \to \mathcal{F} \otimes E_0$ in $Coh_M(Z)$. Then $\mathcal{K}_1, \mathcal{K}_0 \in Coh_M(Z; Z')$. Thus, $\mathcal{K}_1, \mathcal{K}_0$ give rise to elements $\tilde{\mathcal{K}}_1, \tilde{\mathcal{K}}_0 \in K_M(Z; Z')$. Then $\mathcal{F} \mapsto \tilde{\mathcal{K}}_0 - \tilde{\mathcal{K}}_1$ is a well defined homomorphism

$$\gamma_\phi : K_M(Z) \to K_M(Z; Z') = K_M(Z').$$

2.3

Let G be a connected reductive algbraic group over $\mathbb{C}$. Let $\mathfrak{g}$ be the Lie algebra of G and let $\mathfrak{g}_n$ be the variety of nilpotent elements in $\mathfrak{g}$. Let $\mathcal{B}$ be the variety of all Borel subalgebras of $\mathfrak{g}$.

For any parabolic subalgebra $\mathfrak{p}$ of $\mathfrak{g}$ we denote by $\mathfrak{n}_\mathfrak{p}$ the nil-radical of $\mathfrak{p}$.

A parabolic subalgebra $\mathfrak{p}$ of $\mathfrak{g}$ is said to be *almost minimal* if the variety of Borel subalgebras contained in $\mathfrak{p}$ is 1-dimensional. Let I be a finite set indexing the G-orbits on the set of almost minimal parabolic subalgebras (for the adjoint action). A parabolic subalgebra in the G-orbit indexed by i is said to have type i. Let $\mathcal{P}_i$ be the variety of all parabolic subalgebras of type i. Let $\pi_i : \mathcal{B} \to \mathcal{P}_i$ be the morphism defined by $\pi_i(\mathfrak{b}) = \mathfrak{p}$ where $\mathfrak{b} \in \mathcal{B}, \mathfrak{p} \in \mathcal{P}_i, \mathfrak{b} \subset \mathfrak{p}$.

Let $\mathbf{X}$ be the set of isomorphism classes of algebraic G-equivariant line bundles on $\mathcal{B}$ where G acts on $\mathcal{B}$ by the adjoint action. Then $\mathbf{X}$ is a finitely generated free abelian group under the operation given by tensor product of line bundles. For each $i \in I$, let $L_i \in \mathbf{X}$ be the tangent bundle along the fibres of $\pi_i : \mathcal{B} \to \mathcal{P}_i$.

Given $i \in I$ and $L \in \mathbf{X}$, we define an integer m by the requirement that the Euler characteristic of any fibre of π_i (a projective line) with coefficients in the restriction of L to that fibre (regarded as a coherent sheaf) is $m + 1$. We set $m = \check{\alpha}_i(L) \in \mathbb{Z}$. Then $\check{\alpha}_i : \mathbf{X} \to \mathbb{Z}$ is a group homomorphism.

We will often write the tensor product of two line bundles L, L' as LL' and the dual of L as L^{-1}.

Let X be a free abelian group (in additive notation) with a given isomorphism $X \xrightarrow{\sim} \mathbf{X}$ denoted by $x \mapsto L_x$. (Thus, $L_x L_{x'} = L_{x+x'}$ for $x, x' \in X$.) Let

$\alpha_i \in X$ be defined by $L_{\alpha_i} = L_i$. The composition $X \to \mathbf{X} \xrightarrow{\check{\alpha}_i} \mathbb{Z}$ is denoted again by $\check{\alpha}_i$. Let $Y = \mathrm{Hom}(X, \mathbb{Z})$. Then $(X, Y, \alpha_i \in X, \check{\alpha}_i \in Y)$ are as in 1.1 and the most general $(X, Y, \alpha_i \in X, \check{\alpha}_i \in Y)$ in 1.1 is thus obtained. We define $W, \mathcal{H}$ as in 1.1.

2.4

Let $\mathcal{G} = G \times \mathbb{C}^*$. We regard $\mathcal{B}$ and $\mathfrak{g}$ as $\mathcal{G}$-varieties with $\mathcal{G}$-action
$$(g, \lambda) : \mathfrak{b} \mapsto Ad(g)\mathfrak{b} \text{ and } (g, \lambda) : y \mapsto \lambda^{-2} Ad(g)y.$$
Then $\mathfrak{g}_n$ is a $\mathcal{G}$-stable subvariety of $\mathfrak{g}$. Let M be a closed subgroup of $\mathcal{G}$. Then $\mathcal{B}, \mathfrak{g}_n$ can be regarded as M-varieties. We denote by v the element $M \xrightarrow{pr_2} \mathbb{C}^*$ of R_M. Then for any M-variety U, we can regard v as a line bundle in $Vec_M(U)$, by taking the inverse image of $v \in K_M(\text{point})$ under $U \to \text{point}$. For an integer n we define $v^n \in Vec_M(U)$ to be the n-th tensor power of v.

Let $e \in \mathfrak{g}_n$; assume that M is contained in the stabilizer of e in $\mathcal{G}$. Then
$$\mathcal{B}_e = \{\mathfrak{b} \in \mathcal{B} | e \in \mathfrak{b}\}$$
is an M-stable subvariety of $\mathcal{B}$. Let $i \in I$. Let
$$\mathcal{B}_{e,i} = \{\mathfrak{p} \in \mathcal{P}_i | e \in \mathfrak{p}\}, \quad \mathcal{B}'_{e,i} = \{\mathfrak{b} \in \mathcal{B} | e \in \pi_i(\mathfrak{b})\}.$$

Define $a : \mathcal{B}_e \to \mathcal{B}_{e,i}, b : \mathcal{B}'_{e,i} \to \mathcal{B}_{e,i}$ by $\mathfrak{b} \mapsto \pi_i(\mathfrak{b})$. Then a is proper and b is smooth (a P^1-bundle).

Let L be the line bundle on $\mathcal{B}'_{e,i}$ whose fibre at $\mathfrak{b}$ is $\mathfrak{p}/\mathfrak{b}$ where $\mathfrak{p} = \pi_i(\mathfrak{b}) \in \mathcal{P}_i$. This is restriction of the G-equivariant (hence $\mathcal{G}$-equivariant, with trivial $\mathbb{C}^*$-action) line bundle L_i on $\mathcal{B}$ hence is an M-equivariant line bundle. It has a canonical section whose value at $\mathfrak{b}$ is the image of $e \in \pi_i(\mathfrak{b})$ in $\pi_i(\mathfrak{b})/\mathfrak{b}$. Since the sections of L are the same as the sections of $v^{-2}L$, we obtain a section of the line bundle $v^{-2}L$, which is in fact M-invariant. This section of $v^{-2}L$ vanishes exactly over $\mathcal{B}_e$. It defines a map of line bundles $\mathbb{C} \to v^{-2}L$; taking duals and tensoring by v^{-1} we find a map of line bundles $\phi : vL^{-1} \to v^{-1}$ which is an isomorphism outside $\mathcal{B}_e$. By the construction in 2.2, ϕ gives rise to a homomorphism
$$\gamma_\phi : K_M(\mathcal{B}'_{e,i}) \to K_M(\mathcal{B}_e).$$
We define a homomorphism $\mathbf{c}_i : K_M(\mathcal{B}_e) \to K_M(\mathcal{B}_e)$ as the composition
$$K_M(\mathcal{B}_e) \xrightarrow{a_*} K_M(\mathcal{B}_{e,i}) \xrightarrow{b^*} K_M(\mathcal{B}'_{e,i}) \xrightarrow{\gamma_\phi} K_M(\mathcal{B}_e).$$

Next, let $x \in X$. We regard L_x as an object of $Vec_{\mathcal{G}}(\mathcal{B})$ with $\mathbb{C}^*$ acting trivially. We can also regard L_x as an object of $Vec_M(\mathcal{B}_e)$ by restriction from $\mathcal{B}$. Let
$$\theta_x : K_M(\mathcal{B}_e) \to K_M(\mathcal{B}_e) \tag{a}$$
be the homomorphism defined by $\mathcal{F} \mapsto L_x \otimes \mathcal{F}$.

Let $(s,c) \in \mathcal{G}$ be an element in the stabilizer of e and let M be the smallest diagonalizable subgroup of $\mathcal{G}$ containing (s,c). Now (s,c) defines a ring homomorphism $h : R_M \to \mathbb{C}$ (it attaches to an M-module the trace of (s,c) on that M-module). This makes $\mathbb{C}$ into an R_M-module and we can form $\mathbb{C} \otimes_{R_M} K_M(\mathcal{B}_e)$. The operators $\mathbf{c}_i, \theta_x : K_M(\mathcal{B}_e) \to K_M(\mathcal{B}_e)$ are R_M-linear hence they induce operators $\mathbb{C} \otimes_{R_M} K_M(\mathcal{B}_e) \to \mathbb{C} \otimes_{R_M} K_M(\mathcal{B}_e)$ denoted again by $\mathbf{c}_i, \theta_x$.

In the remainder of this section and in the next section we assume that G has simply connected derived group.

Proposition 2.5 *There is a unique $\mathcal{H}$-module structure on $\mathbb{C} \otimes_{R_M} K_M(\mathcal{B}_e)$ such that $v^n \in \mathcal{A}$ acts as multiplication by $h(v^n)$, $v - \tilde{T}_i$ acts as $\mathbf{c}_i$ and θ_x is induced by the map 2.4(a).*

This is proved in [KL] except that there we use equivariant topological K-homology instead of the Grothendieck group $K_M()$. But in our case the two theories coincide, thanks to [DLP].

2.6

Let A be the subgroup of G consisting of all $g \in G$ such that $Ad(g)e = e$ and $gs = sg$. This group acts naturally on $\mathcal{B}_e$ and its action commutes with the action of M. For any $g \in A$ and any $\mathcal{F} \in Coh_M(\mathcal{B}_e)$, we have $(g^{-1})^* \mathcal{F} \in Coh_M(\mathcal{B}_e)$. This defines an action of A on $K_M(\mathcal{B}_e)$ which is R_M-linear hence defines an action of A on $\mathbb{C} \otimes_{R_M} K_M(\mathcal{B}_e)$. For any irreducible $\mathbb{C}$-representation ρ of A which is trivial on the identity component A^0, we consider $E_\rho = \mathrm{Hom}_A(\rho, \mathbb{C} \otimes_{R_M} K_M(\mathcal{B}_e))$. The $\mathcal{H}$-module structure on $\mathbb{C} \otimes_{R_M} K_M(\mathcal{B}_e)$ is compatible with the A-module structure, hence it induces an $\mathcal{H}$-module structure on E_ρ.

Proposition 2.7 *Assume that $c \in \mathbb{C}^*$ is not a root of 1.*

(a) We have $E_\rho \neq 0$ if and only if ρ appears in the complex homology space of $\{\mathfrak{b} \in \mathcal{B}_e | Ad(s)\mathfrak{b} = \mathfrak{b}\}$ regarded as an A-module in a natural way.

(b) If $E_\rho \neq 0$ then E_ρ has a unique simple quotient $\mathcal{H}$-module $\bar{E}_\rho$.

(c) Consider the set of G-conjugacy classes of triples (e, s, ρ) where $e \in \mathfrak{g}_n$, s is a semisimple element of G such that $Ad(s)e = c^2 e$ and ρ is an irreducible representation of A/A^0 (as in 2.6) satisfying (a). On the other hand, consider the set of isomorphism classes of simple $\mathcal{H}$-modules (over $\mathbb{C}$) in which v acts as multiplication by c. These two sets are in bijection under $(e, s, \rho) \mapsto \bar{E}_\rho$.

The comments after the statement of 2.5 apply here as well.

2.8

Let e, h, f be an $\mathfrak{sl}_2$-triple of $\mathfrak{g}$. Let $\zeta : SL_2 \to G$ be the homomorphism of algebraic groups whose tangent map at 1 carries

$\left(\begin{smallmatrix} 0 & 1 \\ 0 & 0 \end{smallmatrix}\right)$ to e, $\left(\begin{smallmatrix} 0 & 0 \\ 1 & 0 \end{smallmatrix}\right)$ to f, $\left(\begin{smallmatrix} 1 & 0 \\ 0 & -1 \end{smallmatrix}\right)$ to h.

Let C be a maximal torus of $\{g \in G | Ad(g)e = e, Ad(g)h = h, Ad(g)f = f\}$. Let $H = C \times \mathbb{C}^*$. We identify H with a closed subgroup of $\mathcal{G}$ via

$$(g, c) \mapsto (g\zeta \left(\begin{smallmatrix} c & 0 \\ 0 & c^{-1} \end{smallmatrix}\right), c).$$

Then $\mathcal{B}_e$ is an H-stable subvariety of $\mathcal{B}$.

Proposition 2.9 *There is a unique $\mathcal{H}$-module structure on $K_H(\mathcal{B}_e)$ such that $v^n \in \mathcal{A}$ acts as multiplication by $v^n \in R_H$, $v - \tilde{T}_i$ acts as $\mathbf{c_i}$ and θ_x acts as in 2.4(a).*

This follows by applying 2.5 for all $(s, c) \in H$ and using [[L5], 1.14].

2.10

Let $i \in I$ and let R be a closed H-stable subvariety of $\mathcal{B}_e$. We say that R is *i-saturated* if the following holds:

$\mathfrak{b} \in R, \mathfrak{b}' \in \mathcal{B}_e, \pi_i(\mathfrak{b}) = \pi_i(\mathfrak{b}') \implies \mathfrak{b}' \in R$.

If this condition is satisfied, then the definition of $\tilde{T}_i$ makes sense for R instead of $\mathcal{B}_e$ and yields an R_H-linear map $\tilde{T}_i : K_H(R) \to K_H(R)$. This is compatible with $\tilde{T}_i : K_H(\mathcal{B}_e) \to K_H(\mathcal{B}_e)$ under the direct image map $K_H(R) \to K_H(\mathcal{B}_e)$ induced by the inclusion $R \subset \mathcal{B}_e$. Hence the image of this map is is stable under $\tilde{T}_i : K_H(\mathcal{B}_e) \to K_H(\mathcal{B}_e)$.

3 Convolution

3.1

In this section we describe an alternative method to define an $\mathcal{H}$-module structure on $K^M(\mathcal{B}_e)$ (as in 2.4) which follows Ginzburg [GI] (with a different normalization, as in [L4]) and is based on a construction (convolution) which first appeared (in a non-equivariant setting) in [KT].

3.2

As a preparation, consider a linear algebraic group M, a smooth M-variety Z and two closed M-stable subsets Z', Z'' of Z. Then we have an R_M-bilinear intersection product (or "Tor product")

$$K_M(Z') \times K_M(Z'') \to K_M(Z' \cap Z'') \tag{a}$$

denoted by $\mathcal{F}', \mathcal{F}'' \mapsto \mathcal{F}' \otimes^L \mathcal{F}''$. It is defined as follows. Start with $\mathcal{F}' \in Coh_M(Z'), \mathcal{F}'' \in Coh_M(Z'')$. We regard $\mathcal{F}', \mathcal{F}''$ as sheaves on Z, zero outside Z', Z'' respectively. We can find a finite resolution of $\mathcal{F}'$ by sheaves in $Vec_M(Z)$; the same applies to Z''. We take the tensor product of these two resolutions and we get a complex of sheaves in $Vec_M(Z)$ which is exact outside $Z' \cap Z''$. The cohomology sheaves of this complex belong to $Coh_M(Z; Z' \cap Z'')$ hence they define elements of $K_M(Z; Z' \cap Z'') = K_M(Z' \cap Z'')$. The alternating sum of these elements is, by definition, $\mathcal{F}' \otimes^L \mathcal{F}''$.

3.3

We preserve the setup of 2.3. Let $i \in I$. Let

$$\bar{O}_i = \{(\mathfrak{b}, \mathfrak{b}') \in \mathcal{B} \times \mathcal{B} | \pi_i(\mathfrak{b}) = \pi_i(\mathfrak{b}')\}.$$

(a closed subvariety of $\mathcal{B} \times \mathcal{B}$). Let $\bar{Z}_i$ be the set of all $(y, \mathfrak{b}, \mathfrak{b}') \in \mathfrak{g}_n \times \mathcal{B} \times \mathcal{B}$ such that $(\mathfrak{b}, \mathfrak{b}') \in \bar{O}_i$ and $y \in \mathfrak{n}_\mathfrak{p}$ where $\mathfrak{p} = \pi_i(\mathfrak{b}) = \pi_i(\mathfrak{b}')$.

We choose $x', x'' \in X$ such that $\check{\alpha}_i(x') = \check{\alpha}_i(x'') = -1$ and $x' + x'' = -\alpha_i$. Then the restriction of $L_{x'} \boxtimes L_{x''}$ from $\mathcal{B} \times \mathcal{B}$ to $\bar{O}_i$ is independent of the choice of x', x''. This restriction, denoted $\mathcal{L}_i$, belongs to $Vec_{\mathcal{G}}(\bar{O}_i)$. The inverse image of $\mathcal{L}_i$ under the obvious map $\bar{Z}_i \to \bar{O}_i$ is again denoted by $\mathcal{L}_i$; it belongs to $Vec_{\mathcal{G}}(\bar{Z}_i)$.

Let $\Lambda = \{(y, \mathfrak{b}) \in \mathfrak{g}_n \times \mathcal{B} | y \in \mathfrak{b}\}$ (a smooth $\mathcal{G}$-variety hence an M-variety). Let $M, e, \mathcal{B}_e$ be as in 2.4. Then $\mathcal{B}_e$ may be regarded as a subvariety of Λ by $\mathfrak{b} \mapsto (e, \mathfrak{b})$. We regard $\bar{Z}_i, \Lambda \times \mathcal{B}_e$ as closed M-stable subvarieties of $\Lambda \times \Lambda$ in an obvious way. Their intersection is $\bar{Z}_i \cap (\mathcal{B}_e \times \mathcal{B}_e)$. Then we have a Tor-product (see 3.2):

$$\otimes^L : K_M(\bar{Z}_i) \times K_M(\Lambda \times \mathcal{B}_e) \to K_M(\bar{Z}_i \cap (\mathcal{B}_e \times \mathcal{B}_e)).$$

The second projection $pr_2 : \Lambda \times \mathcal{B}_e \to \mathcal{B}_e$ is smooth and the first projection $pr_1 : \bar{Z}_i \cap (\mathcal{B}_e \times \mathcal{B}_e) \to \mathcal{B}_e$ is proper. Hence

$$F \mapsto pr_{1*}((v^{-1}\mathcal{L}_i) \otimes^L (pr_2^* F))$$

is a well defined R_M-linear map

$$\mathbf{c}_1' : K_M(\mathcal{B}_e) \to K_M(\mathcal{B}_e).$$

Proposition 3.4 *There is a unique $\mathcal{H}$-module structure on $K_M(\mathcal{B}_e)$ such that $v^n \in \mathcal{A}$ acts as multiplication by $v^n \in R_M$, $-v^{-1} - \tilde{T}_i$ acts as $\mathbf{c}_1'$ and θ_x acts as in 2.4(a).*

See [[L4], Sec. 7, 8 and 10.1].

Proposition 3.5 *If M is as in 2.9, then the $\mathcal{H}$-module structures on $K_M(\mathcal{B}_e)$ described in 2.9 and 3.4 coincide.*

We must only show that the action of $\tilde{T}_i$ in the two module structures on $K_H(\mathcal{B}_e)$ coincide. Both $\mathcal{H}$-module structures can be expressed in terms of equivariant topological K-homology. Now in [[KL], Sec. 5] there is an argument by which one can get information about the action of $\tilde{T}_i$ on a K-homology space of $\mathcal{B}_e$ from information about the action of $\tilde{T}_i$ on a K-homology space of $\mathcal{B}$. (In that reference we use this to show that the relations of $\mathcal{H}$ are satisfied.) But the same argument shows that it is enough to verify the coincidence of the two actions of $\tilde{T}_i$ on a K-homology space of $\mathcal{B}$. In this case we use the explicit formulas [[KL], 3.10],[[L4], 7.23]. This completes the proof.

3.6

In the setup of 3.3 we denote by $\mathcal{B}_e(i)$ the closed subset of $\mathcal{B}_e$ consisting of those $\mathfrak{b}$ such that $e \in \mathfrak{n}_{\mathfrak{p}}$ where $\mathfrak{p} = \pi_i(\mathfrak{b})$. The map $pr_1 : \bar{Z}_i \cap (\mathcal{B}_e \times \mathcal{B}_e) \to \mathcal{B}_e$ has image contained in $\mathcal{B}_e(i)$. Hence the image of $\mathbf{c}'_1 : K_M(\mathcal{B}_e) \to K_M(\mathcal{B}_e)$ is contained in the image of $K_M(\mathcal{B}_e(i)) \to K_M(\mathcal{B}_e)$ (direct image map induced by the inclusion $\mathcal{B}_e(i) \subset \mathcal{B}_e$).

3.7

We will need another K-theoretic construction: Serre-Grothendieck duality.

For any smooth connected M-variety Z we denote by $\Omega_Z \in Vec_M(Z)$ the line bundle of top exterior differential forms on Z.

Let Z be an M-variety. We define a group homomorphism

(a) $D_Z : K_M(Z) \to K_M(Z)$

as follows. We choose an M-equivariant closed imbedding of Z into a smooth M-variety $\tilde{Z}$. Let $\mathcal{F} \in Coh_M(Z)$. We can regard $\mathcal{F}$ as a sheaf on $\tilde{Z}$, zero outside Z. We can find a complex of sheaves $\cdots \to F^p \to F^{p+1} \to \ldots$ in $Coh_M(Z)$ such that F^p are zero for $p > 0$ and for $|p|$ large, and the p-th cohomology sheaf is zero except in degree 0 where it is $\mathcal{F}$. This gives rise to a complex of sheaves $\cdots \to \tilde{F}^p \to \tilde{F}^{p+1} \to \ldots$ in $Vec_M(Z)$ where for any connected component $\tilde{Z}_j$ of $\tilde{Z}$ of dimension n we have $\tilde{F}^p|_{\tilde{Z}_j} = \mathrm{Hom}(F^{-n-p}|_{\tilde{Z}_j}, \Omega_{\tilde{Z}_j})$. The cohomology sheaves of this complex belong to $Coh_M(\tilde{Z}; Z)$ hence they give rise to elements of $K_M(\tilde{Z}; Z) = K_M(Z)$. Taking the alternating sum of these elements, we obtain an element of $K_M(Z)$ which is, by definition, $D_Z(\mathcal{F})$. This defines (a); it is well defined, independent of the choices.

4 Subregular case

4.1

In the remainder of this paper we assume that $\mathfrak{g}$ is simple and that G is simply connected. We fix e, h, f as in 2.8 and assume that e is *subregular*, that is, $\mathcal{B}_e$ (see 2.4) is pure of dimension 1. Let C and $H = C \times \mathbb{C}^* \subset G$ be as in 2.8. Let

$$\Lambda_e = \{(y, \mathfrak{b}) \in \mathfrak{g}_n \times \mathcal{B} \,|\, [y - e, f] = 0, y \in \mathfrak{b}\}.$$

Then Λ_e is an H-stable subvariety of $\mathfrak{g}_n \times \mathcal{B}$. The subvariety $\{e\} \times \mathcal{B}_e = \mathcal{B}_e$ of Λ_e is H-stable. It is known that Λ_e is smooth, connected of dimension 2.

In this section we define a subset $\mathbf{B}_{\mathcal{B}_e}^{\pm} \subset \mathbf{K_H}(\mathcal{B}_e)$ following [L4]. (This can be defined also for arbitrary $e \in \mathfrak{g}_n$, see [L5].)

4.2

Let J be a set that indexes the irreducible components of $\mathcal{B}_e$. Let V_j be the irreducible component of $\mathcal{B}_e$ indexed by $j \in J$. It is isomorphic to P^1. We can regard J as the set of vertices of a graph (called also J) in which j, j' are joined if $V_j \cap V_{j'} \neq \emptyset$. (Then this intersection is a single point $p_{j,j'}$.) It is known that J is a Coxeter graph of type A, D or E.

For $j \in J$, V_j is a projective line; in fact it is a fibre of $\pi_i : \mathcal{B} \to \mathcal{P}_i$ for a well defined $i \in I$. Then $j \mapsto i$ is a map $\omega : J \to I$. This map is surjective and its fibres are the orbits of a cyclic group $\mathbb{Z}/d\mathbb{Z}$ acting on J by graph automorphisms.

If G is of type A, D, E, then $J = I$ and $d = 1$; if G is of type C_n, $(n \geq 3)$, then J is of type D_{n+1} and $d = 2$; if G is of type B_n $(n \geq 2)$, then J is of type A_{2n-1} and $d = 2$; if G is of type F_4, then J is of type E_6 and $d = 2$; if G is of type G_2, then J is of type D_4 and $d = 3$.

A Coxeter graph has a canonical involution (called opposition); it is induced by conjugation by the longest element in the corresponding Weyl group.

4.3

An involution ϖ of the Lie algebra $\mathfrak{g}$ is said to be an *opposition* if, on some Cartan subalgebra of $\mathfrak{g}$, ϖ is multiplication by -1.

We can find (cf. [L5]) an opposition $\varpi : \mathfrak{g} \to \mathfrak{g}$ such that
(a) $\varpi(e) = -e, \varpi(h) = h, \varpi(f) = -f$;
(b) $\varpi = -1$ on the Lie algebra of C;
(c) ϖ induces the opposition involution of the Coxeter graph J.
(Note that, by (a), ϖ induces an involution $\varpi : \mathcal{B}_e \to \mathcal{B}_e$ and this in turn induces an involution of the graph J, so that (c) makes sense.)

If G is of type D, E, we have $C = \{1\}$ hence condition (b) can be omitted; also, condition (c) is automatically satisfied. In this case, ϖ is uniquely determined.

If G is of type C_n, $(n \geq 3)$ or F_4 or G_2 then we have $C = \{1\}$ hence condition (b) can be omitted. In this case, ϖ is uniquely determined by (a),(c).

If G is of type A or B_n $(n \geq 2)$ then dim $C = 1$. In this case, condition (c) is automatically satisfied (in the presence of (a),(b)); moreover ϖ is uniquely determined up to conjugation by $Ad(g)$ for some $g \in C$.

Let $\xi \mapsto \xi^\dagger$ be the involution of R_H induced by the automorphism of $H = C \times \mathbb{C}^*$ given by $(g,c) \mapsto (g^{-1},c)$. The involution ϖ of $\mathcal{B}_e$ is compatible with the H-action in the following way:

$$\varpi((g,c)\mathfrak{b}) = (g,c)^\dagger \varpi(\mathfrak{b})$$

for $(g,c) \in H, \mathfrak{b} \in \mathcal{B}_e$. If $\mathcal{F} \in Coh_H(\mathcal{B}_e)$, then the inverse image of $\mathcal{F}$ under ϖ is again naturally an object of $Coh_H(\mathcal{B}_e)$; we denote it by $\varpi^*\mathcal{F}$. This defines an involution ϖ^* of $K_H(\mathcal{B}_e)$ which is R_H-semilinear with respect to the involution $^\dagger : R_H \to R_H$.

4.4

We consider the R_H-bilinear pairing $K_H(\mathcal{B}_e) \times K_H(\Lambda_e) \to R_H$ given by $(F : F') = \pi_*(F \otimes^L F')$. Here the Tor-product is relative to the smooth variety Λ_e and its closed subvarieties $\mathcal{B}_e, \Lambda_e$ with intersection $\mathcal{B}_e$; π is the map from $\mathcal{B}_e$ to the point.

Let $\nu = l(w_0)$. We now define a pairing $(||) : K_H(\mathcal{B}_e) \times K_H(\Lambda_e) \to R_H$ by

$$(F||F') = (-v)^{\nu-2}(\varpi^*\tilde{T}_{w_0}(F) : F')^\dagger$$

where ϖ^* acts as in 4.3 and $\tilde{T}_{w_0} \in \mathcal{H}$ acts as in 2.9 or 3.4. This pairing is R_H-linear in the first variable and R_H-semilinear (as in 4.3) in the second variable.

We define $\tilde{\beta} : K_H(\mathcal{B}_e) \to K_H(\mathcal{B}_e)$ by

$$\tilde{\beta}(F) = (-v)^{-\nu}\varpi^*\tilde{T}_{w_0}^{-1}D_{\mathcal{B}_e}(F)$$

where ϖ^* acts as in 4.3, $\tilde{T}_{w_0} \in \mathcal{H}$ acts as in 2.9 or 3.4 and $D_{\mathcal{B}_e} : K_H(\mathcal{B}_e) \to K_H(\mathcal{B}_e)$ is as in 3.7. By [[L4], 12.10], $\tilde{\beta}$ is an involution.

Let $\hat{C}$ be the group of characters of C. We have

$$\hat{C} \subset R_C \subset R_C[v, v^{-1}] = R_{C \times \mathbb{C}^*} = R_H.$$

Hence for $\tau \in \hat{C}$ and $F \in K_H(\mathcal{B}_e)$, the product $\tau F \in K_H(\mathcal{B}_e)$ is well defined. From the definition, we see that

$$\tilde{\beta}(\tau F) = \tau\tilde{\beta}(F)$$

for $\tau \in \hat{C}, F \in K_H(\mathcal{B}_e)$.

4.5

Let $k : \mathcal{B}_e \to \Lambda_e$ be the inclusion. As in [[L5], 5.11] we define

$$\mathbf{B}_{\mathcal{B}_e}^\pm = \{F \in K_H(\mathcal{B}_e)|\tilde{\beta}(F) = F, (F||k_*F) \in 1 + R_C[v^{-1}]\}.$$

A priori we only have $(F||k_*F) \in R_H = R_{C \times \mathbb{C}^*} = R_C[v, v^{-1}]$.

4.6

If q is an H-fixed point of V_j and $k : \{q\} \to V_j$, $k' : \{q\} \to \mathcal{B}_e$ are the inclusions, we shall sometimes denote $k_*(\mathbb{C}) \in K_H(V_j)$ and $k'_*(\mathbb{C}) \in K_H(\mathcal{B}_e)$ again by q.

Also, if $O \in Vec_H(V_j)$ we shall denote again by O the object of $Coh_H(\mathcal{B}_e)$ obtained by extending O to $\mathcal{B}_e$ by 0 outside V_j.

4.7

As in [[L6], 3.4] we see that the R_H-module $K_H(\mathcal{B}_e)$ is free. As a basis we may take the elements $O_j^{-1}(j \in J)$ and q; here O_j^{-1} is any H-equivariant line bundle on V_j such that V_j has Euler characteristic 0 with coefficients in O_j^{-1} and q is an H-fixed point in $\mathcal{B}_e$.

From the definition of the action of $\tilde{T}_i$ in 2.9 we see that, if $i = \omega(j)$ and p is an H-fixed point in V_j then

(a) $\tilde{T}_i(O_j^{-1}) = vO_j^{-1}$

and

(b) $\tilde{T}_i(p) = -v^{-1}p + bO_j^{-1}$

for some $b \in R_H$ such that $b \mapsto 0$ under the obvious ring homomorphism $R_H \to R_{\{1\}} = \mathbb{Z}$.

In particular, $\{\xi \in K_H(V_j) | (\tilde{T}_i - v)\xi = 0\} = R_H O_j^{-1}$.

From 3.6 we see that $(\tilde{T}_i + v^{-1})K_H(\mathcal{B}_e)$ is contained in the image of

$$K_H(\cup_{j \in J; \omega(i)=j} V_j) \to K_H(\mathcal{B}_e)$$

(direct image map). Moreover $(\tilde{T}_i + v^{-1})K_H(\mathcal{B}_e)$ must be annihilated by $\tilde{T}_i - v$. Note that $\cup_{j \in J; \omega(i)=j} V_j$ is an i-saturated subvariety of $\mathcal{B}_e$ (a disjoint union of projective lines), see 2.10. It follows that

$$(\tilde{T}_i + v^{-1})K_H(\mathcal{B}_e) \subset \oplus_{j \in J; i=\omega(j)} R_H O_j^{-1}. \tag{c}$$

If $i \in I, j \in J$ are such that $i \neq \omega(j)$, then

$$\tilde{T}_i(O_j^{-1}) = -v^{-1}O_j^{-1} + \sum_{j'} c_{j'} O_{j'}^{-1} \text{ with } c_{j'} \in R_H;$$

here j' runs over the elements of J that are joined with j and satisfy $\omega(j') = i$. (We use (c) and the fact that $V_j \cup (\cup_{j'} V_{j'})$ is i-saturated.)

If $i \in I$ and p is an H-fixed point in $\mathcal{B}_e - \cup_{j \in J; i=\omega(j)} V_j$ then, using (c) and the fact that $\{p\}$ is i-saturated, we see that

(e) $\tilde{T}_i(p) = -v^{-1}p$.

5 Subregular case: type A

5.1

In this section we assume that $G = SL(E)$ where E is a $\mathbb{C}$-vector space with basis $\epsilon_1, \epsilon_2, \ldots, \epsilon_n$ ($n \geq 3$). We set $\epsilon_0 = 0$. We may assume that $e, h, f : E \to E$ are the linear maps

$e\epsilon_k = (n - 1 - k)\epsilon_{k+1}$ for $k \in [1, n - 1]$, $e\epsilon_n = 0$,
$h\epsilon_k = (-n + 2k)\epsilon_k$ for $k \in [1, n - 1]$, $h\epsilon_n = 0$,
$f\epsilon_k = (k - 1)\epsilon_{k-1}$ for $k \in [1, n - 1]$, $f\epsilon_n = 0$.

We identify $\mathcal{B}$ with the variety of complete flags in E in an obvious way. For $k \in [1, n - 1]$, the subspaces

$$E_k = \mathbb{C}\epsilon_{n-k} + \cdots + \mathbb{C}\epsilon_{n-1}, \quad E'_k = \mathbb{C}\epsilon_{n-k+1} + \cdots + \mathbb{C}\epsilon_n$$

are k-dimensional and e-stable. We set $E_0 = 0, E'_n = E$. Note that $e = 0$ on

$$E'_2/E_0, E'_3/E_1, \ldots, E'_n/E_{n-2}.$$

Hence for $k \in [1, n - 1]$, the set V_k of complete flags in E of the form

$$\{E_1 \subset \ldots \subset E_{k-1} \subset W_k \subset E'_{k+1} \subset \ldots \subset E'_{n-1}\}$$

(with variable W_k of dimension k) is a projective line contained in $\mathcal{B}_e$. We have $\mathcal{B}_e = \cup_{k \in [1, n-1]} V_k$. We may take $I = J = [1, n - 1]$. We may assume that C is identified with $\mathbb{C}^*$ and is such that the imbedding of $H = C \times \mathbb{C}^* = \mathbb{C}^* \times \mathbb{C}^*$ in $G = SL(E)$ is given by the following action of $\mathbb{C}^* \times \mathbb{C}^*$ on E:

$$(c', c) : \epsilon_k \to c'^{-1} c^{-n+2k} \epsilon_k \text{ for } k \in [1, n - 1], \quad (c', c) : \epsilon_n \to c'^{n-1} \epsilon_n.$$

This is compatible with the action of e in the following way:

$$(c', c)(ex) = c^2 e(c', c)(x) \text{ for } x \in E.$$

In particular, if $\xi \in \mathcal{B}_e$, then $(c', c)\xi \in \mathcal{B}_e$. Each subspace E_k, E'_k is stable under this action. The fixed point set of the H-action on $\mathcal{B}_e$ consists of the n points:

$$p_{k-1,k} = (E_1 \subset \ldots \subset E_{k-1} \subset E'_k \subset \ldots \subset E'_{n-1}), \quad k \in [1, n].$$

For $k \in [2, n - 1]$ we have $V_{k-1} \cap V_k = \{p_{k-1,k}\}$. Hence in this case, the definition of $p_{k-1,k}$ given above agrees with the definition given in 4.2.

For $k \in [1, n]$, let $L_{k-1,k}$ be the line bundle on the flag manifold whose fibre at $D_* = (0 = D_0 \subset D_1 \subset D_2 \subset \ldots \subset D_{n-1} \subset D_n = E)$ is D_k/D_{k-1}. This is $SL(V)$-equivariant (even $GL(V)$-equivariant). We have $L_{0,1} \otimes L_{1,2} \otimes \ldots \otimes L_{n-1,n} = \mathbb{C}$ as $SL(V)$-equivariant line bundles.

$\check{\alpha}_k$ maps

$$L_{k-1,k} \to -1, \quad L_{k,k+1} \to 1, \quad L_{l-1,l} \to 0, \text{ if } l \neq k, k + 1.$$

5.2

The fibre of $L_{k-1,k}$ at $p_{l-1,l}$ is (as a representation of H):

$E_k/E_{k-1} = v'^{-1}v^{n-2k}$ if $k < l$;

$E'_k/E_{k-1} = v'^{n-1}$ if $k = l$;

$E'_k/E'_{k-1} = v'^{-1}v^{n-2k+2}$ if $k > l$.

Here v' (resp, v) is the character $\mathbb{C}^* \times \mathbb{C}^* \to \mathbb{C}^*$ given by the first (resp. second) projection.

5.3

For $k \in [1, n-1]$, the simple root α_k is $\mathcal{L}_k = L_{k-1,k}^{-1} \otimes L_{k,k+1}$; this is the line bundle on the flag manifold whose fibre at D_* is $\mathrm{Hom}(D_k/D_{k-1}, D_{k+1}/D_k)$. Note that $\mathcal{L}_k|_{V_k}$ is the tangent bundle of V_k. From 5.2 we see that

the fibre of $\mathcal{L}_k$ at $p_{k,k+1}$ is $v'^n v^{-n+2k}$,

the fibre of $\mathcal{L}_k$ at $p_{k-1,k}$ is $v'^{-n} v^{n-2k}$.

5.4

For $k \in [1, n-1]$ let $O = O_k^{b',b;a',a}$ be the H-equivariant line bundle on V_k whose fibre at $p_{k-1,k}$ is $v'^{b'} v^b$ and whose fibre at $p_{k,k+1}$ is $v'^{a'} v^a$. Here $a' - b' = mn$, $a - b = m(-n + 2k)$ where $m + 1$ is the Euler characteristic of V_k with coefficients in O.

If $j : \{p_{k-1,k}\} \to V_k$, $j' : \{p_{k,k+1}\} \to V_k$ are the inclusions, we have exact sequences

$$0 \to O_k^{n,-n+2k;0,0} \to O_k^{0,0;0,0} \to j_*(\mathbb{C}) \to 0,$$

$$0 \to O_k^{0,0;-n,n-2k} \to O_k^{0,0;0,0} \to j'_*(\mathbb{C}) \to 0$$

in $Coh_H(V_k)$. Hence

$$p_{k-1,k} = O_k^{0,0;0,0} - O_k^{n,-n+2k;0,0}, \quad p_{k,k+1} = O_k^{0,0;0,0} - O_k^{0,0;-n,n-2k}$$

in $K_H(V_k)$ and in $K_H(\mathcal{B}_e)$. We have

$$O_k^{b',b;a',a} + O_k^{a',a;b',b} = v'^{a'} v^a + v'^{b'} v^b$$

in $K_H(V_k)$ and in $K_H(\mathcal{B}_e)$, whenever $a' - b' = mn$, $a - b = m(-n + 2k)$ for some m.

5.5

From 5.2, 5.4, we see that the restriction of $L_{k-1,k}$ to V_l is

$O_l^{-1,n-2k;-1,n-2k}$ if $k \in [1, l-1]$,

$O_l^{n-1,0;-1,n-2k}$ if $k = l$,

$O_l^{-1,n-2k+2;n-1,0}$ if $k = l+1$,

$O_l^{-1,n-2k+2;-1,n-2k+2}$ if $k \in [l+2, n]$.

Lemma 5.6 Let $k \in [1, n-1]$. We have $\tilde{T}_k(O_k^{0,0;-n,n-2k}) = vO_k^{0,0;-n,n-2k}$ in $K_H(V_k)$ or in $K_H(\mathcal{B}_e)$.

This is a special case of 4.7(a).

Lemma 5.7 Let $k \in [1, n-1]$ and let $\xi \in K_H(\mathcal{B}_e)$. We have
$L_{k-1,k}(\tilde{T}_k\xi) = (\tilde{T}_k + v^{-1} - v)(L_{k,k+1}\xi)$.

Here $L_{k-1,k}$ acts as by tensor product. This is a special case of the relation 1.1(e) which holds in the $\mathcal{H}$-module $K_H(\mathcal{B}_e)$.

Lemma 5.8 (a) $\tilde{T}_1(p_{01}) = -v^{-1}p_{01} - O_1^{n,-n+1;0,-1} + O_1^{0,1;-n,n-1}$.
(b) $\tilde{T}_k(p_{01}) = -v^{-1}p_{01}$ for $k \geq 2$.

We prove (a). By 4.7(b), we have
$\tilde{T}_1(p_{01}) = -v^{-1}p_{01} + bO_1^{0,0;-n,n-2}$ for some $b \in R_H$.
Let $\zeta = O_1^{0,0;-n,n-2}$. By 5.7, we have

$$L_{0,1}\tilde{T}_1 p_{01} = (\tilde{T}_1 + v^{-1} - v)(L_{0,1}p_{01}) = v'^{-1}v^{n-2}(\tilde{T}_1 + v^{-1} - v)p_{01},$$

$$L_{0,1}(-v^{-1}p_{01} + bO_1^{0,0;-n,n-2}) =$$
$$= v'^{-1}v^{n-2}(-v^{-1}p_{01} + bO_1^{0,0;-n,n-2} + (v^{-1} - v)p_{01}),$$

$$-v^{-1}v'^{n-1}p_{01} + bO_1^{n-1,0;-1,n-2}O_1^{0,0;-n,n-2} = -v'^{-1}v^{n-1}p_{01} \quad \text{mod } \zeta.$$

The last equality takes place in $K_H(V_1)$. Now

$$O_1^{n-1,0;-1,n-2}O_1^{0,0;-n,n-2} = O_1^{n-1,0;-1,n-2}(-O_1^{-n,n-2;0,0} + v'-nv^{n-2} + 1) =$$

$$= -O_1^{-1,n-2;-1,n-2} \quad \text{mod } \zeta = -v'^{-1}v^{n-2}p_{01} \quad \text{mod } \zeta.$$

Hence

$$-v^{-1}v'^{n-1}p_{01} - bv'^{-1}v^{n-2}p_{01} = -v'^{-1}v^{n-1}p_{01} \quad \text{mod } \zeta,$$

$$-v^{-1}v'^{n-1} - bv'^{-1}v^{n-2} = -v'^{-1}v^{n-1}.$$

Thus, $b = -v'^n v^{-n+1} + v$; this proves (a). Now (b) follows from 4.7(e).

Lemma 5.9 *Assume that $k \in [2, n-1]$. We have (in $K_H(\mathcal{B}_e)$):*

(a) $\tilde{T}_k O_{k-1}^{0,0;-n,n-2k+2} = -v^{-1} O_{k-1}^{0,0;-n,n-2k+2} - v O_k^{0,0;-n,n-2k}$

(b) $\tilde{T}_{k-1} O_k^{0,0;-n,n-2k} = -v^{-1} O_k^{0,0;-n,n-2k} - v^{-1} O_{k-1}^{0,0;-n,n-2k+2}$.

We prove (a). Let $\tilde{p} = p_{k-1,k}$. By 4.7(d), we have

$$\tilde{T}_k O_{k-1}^{0,0;-n,n-2k+2} = -v^{-1} O_{k-1}^{0,0;-n,n-2k+2} + c O_k^{0,0;-n,n-2k}$$

for some $c \in R_H$. We have

$$L_{k-1,k} \tilde{T}_k O_{k-1}^{0,0;-n,n-2k+2} = (\tilde{T}_k + v^{-1} - v) L_{k,k+1} O_{k-1}^{0,0;-n,n-2k+2}$$

$$L_{k-1,k}(-v^{-1} O_{k-1}^{0,0;-n,n-2k+2} + c O_k^{0,0;-n,n-2k}) =$$
$$= (\tilde{T}_k + v^{-1} - v) v'^{-1} v^{n-2k} O_{k-1}^{0,0;-n,n-2k+2}$$
$$= v'^{-1} v^{n-2k} (-v^{-1} O_{k-1}^{0,0;-n,n-2k+2} + c O_k^{0,0;-n,n-2k}$$
$$+ (v^{-1} - v) O_{k-1}^{0,0;-n,n-2k+2})$$

$$L_{k-1,k}(-v^{-1} O_{k-1}^{0,0;-n,n-2k+2} + c O_k^{0,0;-n,n-2k}) =$$
$$= (\tilde{T}_k + v^{-1} - v) v'^{-1} v^{n-2k} O_{k-1}^{0,0;-n,n-2k+2}$$
$$= v'^{-1} v^{n-2k} (-v^{-1} O_{k-1}^{0,0;-n,n-2k+2} + c O_k^{0,0;-n,n-2k}$$
$$+ (v^{-1} - v) O_{k-1}^{0,0;-n,n-2k+2}),$$

$$- v^{-1} O_{k-1}^{-1,n-2k+2;n-1,0} O_{k-1}^{0,0;-n,n-2k+2} + c O_k^{n-1,0;-1,n-2k} O_k^{0,0;-n,n-2k} =$$
$$= v'^{-1} v^{n-2k} (c O_k^{0,0;-n,n-2k} - v O_{k-1}^{0,0;-n,n-2k+2}).$$

Here the products are taken in $K_H(V_{k-1})$ or $K_H(V_k)$ (note that V_{k-1}, V_k are smooth). We have

$$O_{k-1}^{-1,n-2k+2;n-1,0} O_{k-1}^{0,0;-n,n-2k+2} = v'^{-1} v^{n-2k+2} O_{k-1}^{0,0;0,0}$$
$$= v'^{-1} v^{n-2k+2} (\tilde{p} + O_{k-1}^{0,0;-n,n-2k+2}),$$
$$O_k^{n-1,0;-1,n-2k} \otimes O_k^{0,0;-n,n-2k} =$$
$$= O_k^{n-1,0;-1,n-2k} (-O_k^{-n,n-2k;0,0} + v'^{-n} v^{n-2k} + 1)$$
$$= -O_k^{-1,n-2k;-1,n-2k} + (v'^{-n} v^{n-2k} + 1) O_k^{n-1,0;-1,n-2k}$$
$$= -v'^{-1} v^{n-2k} (\tilde{p} + O_k^{n,-n+2k;0,0}) + (v'^{-n} v^{n-2k} + 1) O_k^{n-1,0;-1,n-2k}.$$

We get

$$- v'^{-1} v^{n-2k+1} (\tilde{p} + O_{k-1}^{0,0;-n,n-2k+2})$$
$$- c v'^{-1} v^{n-2k} (\tilde{p} + O_k^{n,-n+2k;0,0}) + c(v'^{-n} v^{n-2k} + 1) O_k^{n-1,0;-1,n-2k}$$
$$= v'^{-1} v^{n-2k} (c O_k^{0,0;-n,n-2k} - v O_{k-1}^{0,0;-n,n-2k+2}).$$

The coefficient of $\tilde{p}$ gives $-v'^{-1}v^{n-2k+1} - cv'^{-1}v^{n-2k} = 0$ hence $c = -v$. This proves (a). The proof of (b) is similar.

Lemma 5.10 *Assume that $k, k' \in [1, n-1]$, $k' \neq k, k \pm 1$. Then*
$$\tilde{T}_k(O_{k'}^{0,0;-n,n-2k'}) = -v^{-1}O_{k'}^{0,0;-n,n-2k'}.$$

This is a special case of 4.7(d).

5.11

For $k \in [1, n-1]$ we set
$$\mathbf{O}_k = O_k^{0,-n+k;-n,-k} \in K_H(\mathcal{B}_e).$$

We rewrite some of the earlier identities as follows. Here k, l are such that both sides make sense.
$$\tilde{T}_k\mathbf{O}_{k-1} = -v^{-1}\mathbf{O}_{k-1} - \mathbf{O}_k;$$
$$\tilde{T}_{k-1}\mathbf{O}_k = -v^{-1}\mathbf{O}_k - \mathbf{O}_{k-1};$$
$$\tilde{T}_l(\mathbf{O}_k) = -v^{-1}\mathbf{O}_k \text{ if } l \neq k-1, k, k+1;$$
$$\tilde{T}_k(\mathbf{O}_k) = v\mathbf{O}_k;$$
$$p_{k-1,k} = p_{k,k+1} + (-v'^n v^k + v^{n-k})\mathbf{O}_k;$$
$$\tilde{T}_k(p_{01}) = -v^{-1}p_{01} \text{ for } k \geq 2;$$
$$\tilde{T}_1(p_{01}) = -v^{-1}p_{01} + (-v'^n + v^n)\mathbf{O}_1.$$

5.12

From 5.11 we see that
$$p_{01} = p_{n-1,n} + \sum_{k \in [1,n-1]} (-v'^n v^k + v^{n-k})\mathbf{O}_k.$$

For $k \in [1, n-1]$, let
$$A_k = (v'^n - v^n)(v^{n-k} - v^{-n+k})/(v^n - v^{-n}),$$
$$A'_k = (-v'^n v^n + 1)(v^k - v^{-k})/(v^n - v^{-n}).$$
Then $A'_k = A_k + (-v'^n v^k + v^{n-k})$ hence we can set
$$\xi = p_{01} + \sum_{k \in [1,n-1]} A_k\mathbf{O}_k = p_{n-1,n} + \sum_{k \in [1,n-1]} A'_k\mathbf{O}_k.$$

The equations
$$(v + v^{-1})A_1 = A_2 + v'^n - v^n, \quad (v + v^{-1})A_2 = A_1 + A_3,$$
$$(v + v^{-1})A_3 = A_2 + A_4, \ldots, \quad (v + v^{-1})A_{n-1} = A_{n-2}$$

can be also expressed in the form
$$\tilde{T}_k\xi = -v^{-1}\xi$$

for all $k \in [1, n-1]$.

Lemma 5.13 *For $k \in [1, n-1]$ we have*

(a) $\tilde{T}_{w_0}^{-1} \mathbf{O}_k = -(-1)^\nu v^{\nu-n} \mathbf{O}_{n-k}$,

(b) $\tilde{T}_{w_0} \mathbf{O}_k = -(-1)^\nu v^{-\nu+n} \mathbf{O}_{n-k}$.

(Compare [[L6], 4.5].) Let $\mathcal{H}_0$ be the $\mathcal{A}$-subalgebra of $\mathcal{H}$ generated by $\{\tilde{T}_k | k \in [1, n-1]\}$. Let $\mathcal{M}$ be the R_H-submodule of $K_H(\mathcal{B}_e)$ with basis $\{\mathbf{O}_k | k \in [1, n-1]\}$. Then $\mathcal{M}$ is an $\mathcal{H}_0$-submodule of $K_H(\mathcal{B}_e)$. Since $\{m \in \mathcal{M} | \tilde{T}_k m = vm\} = R_H \mathbf{O}_k\}$ and $\tilde{T}_{w_0} \tilde{T}_k \tilde{T}_{w_0}^{-1} = \tilde{T}_{n-k}$ it follows that $\tilde{T}_{w_0}(R_H \mathbf{O}_k) = R_H \mathbf{O}_{n-k}$. Hence $\tilde{T}_{w_0} \mathbf{O}_k = b_k \mathbf{O}_{n-k}$ where $b_k \in R_H$. Note that b_k is invertible in R_H since $\tilde{T}_{w_0} : \mathcal{M} \to \mathcal{M}$ is an isomorphism.

If $k \in [2, n]$ we have $\tilde{T}_k \mathbf{O}_{k-1} = -v^{-1} \mathbf{O}_{k-1} - \mathbf{O}_k$, hence

$\tilde{T}_{w_0} \tilde{T}_k \mathbf{O}_{k-1} = -v^{-1} \tilde{T}_{w_0} \mathbf{O}_{k-1} - \tilde{T}_{w_0} \mathbf{O}_k$,

$\tilde{T}_{n-k} \tilde{T}_{w_0} \mathbf{O}_{k-1} = b_{k-1} \tilde{T}_{n-k} \mathbf{O}_{k-1} = -v^{-1} b_{k-1} \mathbf{O}_{n-k+1} - b_k \mathbf{O}_{n-k}$,

$\tilde{T}_{n-k} \mathbf{O}_{k-1} = -v^{-1} \mathbf{O}_{n-k+1} - b_k b_{k-1}^{-1} \mathbf{O}_{n-k}$.

It follows that $b_k b_{k-1}^{-1} = 1$. Thus, b_k is independent of k. Thus there exists an invertible element $\epsilon v^{\prime c'} v^c \in R_H$ with $\epsilon = \pm 1$, $c, c' \in \mathbb{Z}$ such that $\tilde{T}_{w_0} \mathbf{O}_k = \epsilon v^{\prime c'} v^c \mathbf{O}_{n-k}$ for all k. The determinant (over R_H) of $\tilde{T}_{w_0} : \mathcal{M} \to \mathcal{M}$ is on the one hand equal to $\pm v^{c(n-1)} v^{\prime c'(n-1)}$ and on the other hand is equal to the ν-th power of the determinant of $\tilde{T}_k : \mathcal{M} \to \mathcal{M}$ where $k \in [1, n-1]$, that is to $((-1)^n v^{-n+3})^\nu$. Thus, $\pm v^{c(n-1)} v^{\prime c'(n-1)} = ((-1)^n v^{-n+3})^\nu$. It follows that $c' = 0$ and $c = (-n+3)n/2$. This proves (b) up to sign. To determine the sign, we specialize $v = 1$. Under this specialization $\mathcal{M}$ becomes the reflection representation of W tensor the sign representation. In this representation w_0 acts as $\mathbf{O}_k \mapsto -(-1)^\nu \mathbf{O}_{n-k}$. This proves (b). Now (a) follows from (b). The lemma is proved.

Lemma 5.14

(a) $\tilde{T}_{w_0}^{-1} p_{n-1,n} = (-v)^\nu p_{01} - (-1)_{k \in [1,n-1]}^\nu v^\nu (v^{n-k} - v^{-n+k}) \mathbf{O}_k$.

(b) $\tilde{T}_{w_0} p_{n-1,n} = (-v)^{-\nu} p_{01} - (-1)_{k \in [1,n-1]}^\nu v^{\prime n} v^{-\nu+n} (v^{n-k} - v^{-n+k}) \mathbf{O}_k$.

We prove (a). Let ξ be as in 5.12. Clearly, $\tilde{T}_{w_0}^{-1}(\xi) = (-v)^\nu \xi$, or equivalently

$$\tilde{T}_{w_0}^{-1}\left(p_{n-1,n} + \sum_{k \in [1,n-1]} A_k' \mathbf{O}_k\right) = (-v)^\nu \left(p_{01} + \sum_k A_k \mathbf{O}_k\right).$$

Hence

$$\tilde{T}_{w_0}^{-1}(p_{n-1,n}) - \sum_k A_{n-k}'(-1)^\nu v^{\nu-n} \mathbf{O}_k = (-v)^\nu p_{01} + (-v)^\nu \sum_k A_k \mathbf{O}_k,$$

$$\tilde{T}_{w_0}^{-1}(p_{n-1,n}) = (-v)^\nu p_{01} + (-1)^\nu \sum_k (v^{\nu-n} A_{n-k}' + v^\nu A_k) \mathbf{O}_k.$$

We have

$$v^{-n} A_{n-k}' + A_k = -(v^{n-k} - v^{-n+k}).$$

This proves (a). We prove (b). Clearly, $\tilde{T}_{w_0}(\xi) = (-v)^{-\nu} \xi$ or equivalently

$$\tilde{T}_{w_0}(p_{n-1,n} + \sum_k A'_k \mathbf{O}_k) = (-v)^{-\nu}(p_{01} + \sum_k A_k \mathbf{O}_k).$$

Hence

$$\tilde{T}_{w_0}(p_{n-1,n}) - \sum_k A'_{n-k}(-1)^\nu v^{-\nu+n} \mathbf{O}_k = (-v)^{-\nu} p_{01} + (-v)^{-\nu} \sum_k A_k \mathbf{O}_k,$$

$$\tilde{T}_{w_0}(p_{n-1,n}) = (-v)^{-\nu} p_{01} + (-1)^\nu \sum_k (v^{-\nu+n} A'_{n-k} + v^{-\nu} A_k) \mathbf{O}_k.$$

We have

$$A'_{n-k} + v^{-n} A_k = -v'^n (v^{n-k} - v^{-n+k}).$$

The lemma is proved.

Lemma 5.15 *For* $k \in [1, n-1]$ *we have* $\varpi^* \mathbf{O}_k = v'^n \mathbf{O}_{n-k}$.

It is enough to show that $\varpi^* O_k^{0,-n+k;-n,-k} = O_{n-k}^{n,-k;0,-n+k}$ as objects of $Vec_H(V_{n-k})$. By definition, the fibre of $\varpi^* O_k^{0,-n+k;-n,-k}$ at $p_{n-k-1,n-k} = \varpi(p_{k,k+1})$ is $v'^n v^{-k}$ and its fibre at $p_{n-k,n-k+1} = \varpi(p_{k-1,k})$ is v^{-n+k}. These are the same as the fibres of $O_{n-k}^{n,-k;0,-n+k}$ at the corresponding points. The lemma follows.

Lemma 5.16 *For* $k \in [1, n-1]$ *we have* $D_{\mathcal{B}_e}(\mathbf{O}_k) = -v'^n v^n \mathbf{O}_k$.

We have $\Omega_{V_k} = O_k^{n,-n+2k;-n,n-2k}$. Hence

$$D_{\mathcal{B}_e}(\mathbf{O}_k) = D_{\mathcal{B}_e}(O_k^{0,-n+k;-n,-k}) = -O_k^{0,n-k;n,k} \Omega_{V_k}$$
$$= -O_k^{0,n-k;n,k} O_k^{n,-n+2k;-n,n-2k} = -O_k^{n,k;0,n-k} = -v'^n v^n \mathbf{O}_k.$$

5.17

We set

$$\mathbf{O}_n = p_{01} - \sum_{k \in [1,n-1]} v^{n-k} \mathbf{O}_k = p_{n-1,n} - \sum_{k \in [1,n-1]} v'^n v^k \mathbf{O}_k.$$

Lemma 5.18 *For* $k \in [1, n]$, *we have* $(-v)^{-\nu} \tilde{T}_{w_0}^{-1} \varpi^* D_{\mathcal{B}_e}(\mathbf{O}_k) = \mathbf{O}_k$.

Assume first that $k \in [1, n-1]$. We have

$$(-v)^{-\nu} \tilde{T}_{w_0}^{-1} \varpi^* D_{\mathcal{B}_e}(\mathbf{O}_k) = (-v)^{-\nu} \tilde{T}_{w_0}^{-1} \varpi^* (-v'^n v^n \mathbf{O}_k) =$$
$$- (-v)^{-\nu} v^n \tilde{T}_{w_0}^{-1} \mathbf{O}_{n-k}$$
$$= (-v)^{-\nu} v^n (-1)^\nu v^{\nu-n} \mathbf{O}_{n-k}.$$

Next we note that $D_{\mathcal{B}_e}(p_{01}) = p_{01}$ and $\varpi^*(p_{01}) = p_{n-1,n}$. We now consider the case $k = n$ using the already known case $k \in [1, n-1]$:

$$(-v)^{-\nu}\tilde{T}_{w_0}^{-1}\varpi^* D_{\mathcal{B}_e}(\mathbf{O}_n) = (-v)^{-\nu}\tilde{T}_{w_0}^{-1}\varpi D(p_{01} - \sum_{k \in [1,n-1]} v^{n-k}\mathbf{O}_k)$$

$$= (-v)^{-\nu}\tilde{T}_{w_0}^{-1}(p_{n-1,n}) - \sum_{k \in [1,n-1]} v^{-n+k}\mathbf{O}_k$$

$$= p_{01} - \sum_{k \in [1,n-1]} (v^{n-k} - v^{-n+k})\mathbf{O}_k - \sum_{k \in [1,n-1]} v^{-n+k}\mathbf{O}_k$$

$$= p_{01} - \sum_{k \in [1,n-1]} v^{n-k}\mathbf{O}_k = \mathbf{O}_n.$$

The lemma is proved.

Lemma 5.19 *We have*

$$\tilde{T}_{w_0}\mathbf{O}_n = (-v)^{-\nu}p_{01} + (-v)^{-\nu}v'^n \sum_{k \in [1,n-1]} v^k\mathbf{O}_k$$

$$= (-v)^{-\nu}p_{n-1,n} + (-v)^{-\nu} \sum_{k \in [1,n-1]} v^{n-k}\mathbf{O}_k.$$

We have

$$\tilde{T}_{w_0}\mathbf{O}_n = \tilde{T}_{w_0}(p_{n-1,n} - \sum_{k \in [1,n-1]} v'^n v^k\mathbf{O}_k)$$

$$= (-v)^{-\nu}p_{01} - (-1)^\nu \sum_{k \in [1,n-1]} v'^n v^{-\nu+n}(v^{n-k} - v^{-n+k})\mathbf{O}_k$$

$$+ (-1)^\nu v^{-\nu+n} \sum_{k \in [1,n-1]} v'^n v^{n-k}\mathbf{O}_k.$$

The lemma follows.

5.20

Consider a hermitian form $(,)$ on $K_H(\mathcal{B}_e)$ with values in R_H (linear in the first variable, antilinear in the second with respect to the ring involution $\dagger : R_H \to R_H$ (that is, $v'^\dagger = v'^{-1}, v^\dagger = v$) such that

$$(\chi\xi, \xi') = (\xi, \chi^\blacktriangle\xi') \text{ and } (\xi, \xi') = (\xi', \xi)^\dagger$$

for $\chi \in \mathcal{H}, \xi, \xi' \in K_H(\mathcal{B}_e)$.

Lemma 5.21 *There exists $c \in R_H$ such that*
(a) $(\mathbf{O}_k, \mathbf{O}_{k'}) = c$ if k, k' in $[1, n-1]$ are consecutive;
(b) $(\mathbf{O}_k, \mathbf{O}_k) = -(v + v^{-1})c$ for all $k \in [1, n-1]$;

(c) $(\mathbf{O}_k, \mathbf{O}_{k'}) = 0$ if $k \neq k'$ in $[1, n-1]$ are not consecutive;

(d) $(p_{01}, \mathbf{O}_k) = 0$ if $k \in [2, n-1]$;

(e) $(p_{01}, \mathbf{O}_1) = -c(-v'^n + v^n)$;

(f) $(p_{01}, p_{01}) = -cv^{-1}(v'^n - v^n)(v'^{-n} - v^n)$.

(a)-(d) are proved as in [[L6], 5.1]. We have $(\tilde{T}_1 p_{01}, \mathbf{O}_1) = (p_{01}, \tilde{T}_1 \mathbf{O}_1)$ hence

$$(-v^{-1}p_{01} + (-v'^n + v^n)\mathbf{O}_1, \mathbf{O}_1) = (p_{01}, v\mathbf{O}_1),$$

$$(v + v^{-1})(p_{01}, \mathbf{O}_1) = -c(-v'^n + v^n)(v + v^{-1})$$

and (e) follows.

We write θ_L instead of θ_x where $L = L_x, x \in X$. From the definitions we have $\theta_{L_{01}^{-1}}^{\blacktriangle} = \tilde{T}_{w_0}^{-1} \theta_{L_{n-1,n}} \tilde{T}_{w_0}$ in $\mathcal{H}$. Hence

$$(\theta_{L_{01}^{-1}} p_{01}, \mathbf{O}_1) = (p_{01}, \tilde{T}_{w_0}^{-1} \theta_{L_{n-1,n}} \tilde{T}_{w_0} \mathbf{O}_1) =$$
$$= -(-1)^{\nu} v^{-\nu+n}(p_{01}, \tilde{T}_{w_0}^{-1} \theta_{L_{n-1,n}} \mathbf{O}_{n-1}).$$

Recall that $L_{n-1,n}|_{V_{n-1}} = O_{n-1}^{-1,-n+2;n-1,0}$ hence

$$\theta_{L_{n-1,n}} \mathbf{O}_{n-1} = O_{n-1}^{-1,-n+2;n-1,0} O_{n-1}^{0,-1;-n,-n+1} = O_{n-1}^{-1,-n+1;-1,-n+1}$$
$$= v'^{-1}v^{-n+1}O_{n-1}^{0,0;0,0} = v'^{-1}v^{-n+1}(p_{n-1,n} + O_{n-1}^{0,0;-n,-n+2})$$
$$= v'^{-1}v^{-n+1}(p_{n-1,n} + v\mathbf{O}_{n-1}).$$

We have $\theta_{L_{01}^{-1}} p_{01} = v'^{-n+1} p_{01}$. Thus,

$$v'^{-n+1}(p_{01}, \mathbf{O}_1)$$
$$= -(-1)^{\nu} v^{-\nu+n}(p_{01}, \tilde{T}_{w_0}^{-1}(v'^{-1}v^{-n+1}p_{n-1,n} + v'^{-1}v^{-n+2}\mathbf{O}_{n-1}))$$
$$= -(-1)^{\nu} v^{-\nu+n}(p_{01}, v'^{-1}v^{-n+1}((-v)^{\nu}p_{01} -$$
$$- (-1)^{\nu} \sum_{k \in [1,n-1]} v^{\nu}(v^{n-k} - v^{-n+k})\mathbf{O}_k) - (-1)^{\nu}v'^{-1}v^{-n+2}v^{\nu-n}\mathbf{O}_1)$$
$$= -v'v(p_{01}, p_{01}) + v'v^n(p_{01}, \mathbf{O}_1).$$

Thus, $v'^{-n+1}(p_{01}, \mathbf{O}_1) = -v'v(p_{01}, p_{01}) + v'v^n(p_{01}, \mathbf{O}_1)$. Here we substitute $(p_{01}, \mathbf{O}_1)$ by the expression in (e); (f) follows. The lemma is proved.

5.22

We now take $(\xi, \xi') = (\xi \| k_* \xi')$ (notation of 5.4, 5.5). This satisfies the conditions in 5.20 (see [L4]) hence 5.21 is applicable.

Lemma 5.23 For this $(,)$ we have $c = -v^{-1}$.

We have

$$(\mathbf{O}_2, \mathbf{O}_1) = (-v)^{\nu-2}(\mathbf{O}_2 : \tilde{T}_{w_0}\varpi^*(\mathbf{O}_1)) = (-v)^{\nu-2}(\mathbf{O}_2 : \tilde{T}_{w_0}v'^n\mathbf{O}_{n-1})$$
$$= (-v)^{\nu-2}v'^n(\mathbf{O}_2 : -(-1)^\nu v^{-\nu+n}\mathbf{O}_1) = -v'^n v^{n-2}(\mathbf{O}_2 : \mathbf{O}_1).$$

The fibre of $\mathbf{O}_2 = O_2^{0,-n+2;-n,-2}$ at p_{12} is v^{-n+2} and the fibre of $\mathbf{O}_1 = O_1^{0,-n+1;-n,-1}$ at p_{12} is $v'^{-n}v^{-1}$. Since $n \geq 3$, V_2, V_1 intersect transversally in Λ_e at p_{12}, so that

$$(\mathbf{O}_2 : \mathbf{O}_1) = v^{-n+2}(v'^{-n}v^{-1}) = v'^{-n}v^{-n+1}.$$

Hence $(\mathbf{O}_2, \mathbf{O}_1) = -v'^n v^{n-2}v'^{-n}v^{-n+1} = -v^{-1}$.
On the other hand, $(\mathbf{O}_2, \mathbf{O}_1) = c$. The lemma is proved.

Lemma 5.24 *We have*
  (a) $(\mathbf{O}_k, \mathbf{O}_{k'}) = -v^{-1}$ *if* k, k' *in* $[1, n-1]$ *are consecutive;*
  (b) $(\mathbf{O}_k, \mathbf{O}_k) = 1 + v^{-2}$ *for all* $k \in [1, n-1]$;
  (c) $(\mathbf{O}_k, \mathbf{O}_{k'}) = 0$ *if* $k \neq k'$ *in* $[1, n-1]$ *are not consecutive;*
  (d) $(p_{01}, \mathbf{O}_k) = 0$ *if* $k \in [2, n-1]$;
  (e) $(p_{01}, \mathbf{O}_1) = v^{-1}(-v'^n + v^n)$;
  (f) $(p_{01}, p_{01}) = v^{-2}(v'^n - v^n)(v'^{-n} - v^n)$;
  (g) $(\mathbf{O}_n, \mathbf{O}_1) = -v^{-1}v'^n$;
  (h) $(\mathbf{O}_n, \mathbf{O}_{n-1}) = -v^{-1}$;
  (i) $(\mathbf{O}_n, \mathbf{O}_k) = 0$ *for* $k \in [2, n-2]$;
  (j) $(\mathbf{O}_n, \mathbf{O}_n) = 1 + v^{-2}$.

(a)-(f) follow from 5.21, 5.23. We prove (g). We have

$$(\mathbf{O}_n, \mathbf{O}_1) = (p_{01} - \sum_{k \in [1, n-1]} v^{n-k}\mathbf{O}_k, \mathbf{O}_1)$$
$$= (p_{01}, \mathbf{O}_1) - v^{n-1}(\mathbf{O}_1, \mathbf{O}_1) - v^{n-2}(\mathbf{O}_2, \mathbf{O}_1)$$
$$= v^{-1}(-v'^n + v^n) - v^{n-1}(1 + v^{-2}) + v^{n-2}v^{-1} = -v^{-1}v'^n.$$

We prove (h). We have

$$(\mathbf{O}_n, \mathbf{O}_{n-1}) = (p_{01} - \sum_{k \in [1, n-1]} v^{n-k}\mathbf{O}_k, \mathbf{O}_{n-1})$$
$$= (p_{01}, \mathbf{O}_{n-1}) - v(\mathbf{O}_{n-1}, \mathbf{O}_{n-1}) - v^2(\mathbf{O}_{n-2}, \mathbf{O}_{n-1})$$
$$= -v(1 + v^{-2}) + v^2 v^{-1} = -v^{-1}.$$

We prove (i). For $k \in [2, n-2]$ we have

$$(\mathbf{O}_n, \mathbf{O}_k) = (p_{01} - \sum_{k' \in [1, n-1]} v^{n-k'}\mathbf{O}_{k'}, \mathbf{O}_k)$$
$$= -v^{n-k}(1 + v^{-2}) + v^{n-k-1}v^{-1} + v^{n-k+1}v^{-1} = 0.$$

We prove (j). We have

$$(\mathbf{O}_n, \mathbf{O}_n) = (p_{01} - \sum_k v^{n-k}\mathbf{O}_k, p_{01} - \sum_k v^{n-k}\mathbf{O}_k)$$

$$= v^{-2}(v'^n - v^n)(v'^{-n} - v^n) - v^{-1}v^{n-1}(-v'^n + v^n) - v^{-1}v^{n-1}(-v'^{-n} + v^n)$$

$$+ \sum_{k\in[1,n-1]} v^{2k}(1 + v^{-2}) - 2 \sum_{k\in[1,n-2]} v^{2k+1}v^{-1} = 1 + v^{-2}.$$

The lemma is proved.

Proposition 5.25 $\mathbf{B}_{\mathcal{B}_e}^{\pm}$ *is the signed $\mathcal{A}$-basis of $K_H(\mathcal{B}_e)$ consists of the elements $\pm v'^s\mathbf{O}_k(k \in [1,n], s \in \mathbb{Z})$.*

Any element ξ in the list above satisfies $(\xi, \xi) = 1 + v^{-2}$ by 5.24 and $\tilde{\beta}(\xi) = \xi$ by 5.18. Hence $\xi \in \mathbf{B}_{\mathcal{B}_e}^{\pm}$.

Conversely, let $\xi' \in \mathbf{B}_{\mathcal{B}_e}^{\pm}$. Let $\partial : R_H \to \mathcal{A}$ be the group homomorphism given by $v'^s v^t \mapsto v^t$ if $s = 0$ and $v'^s v^t \mapsto 0$ if $s \neq 0$. Now the elements $\mathbf{O}_k(k \in [1, n-1])$ and p_{01} form an R_H-basis of $K_H(\mathcal{B}_e)$. Using the definition of $\mathbf{O}_n$ (see 5.17) we deduce that the elements $\mathbf{O}_k(k \in [1,n])$ form an R_H-basis of $K_H(\mathcal{B}_e)$. Thus, we have $\xi' = \sum_{s\in\mathbb{Z},k\in[1,n]} c_{s,k}v'^s\mathbf{O}_k$ where $c_{s,k} \in \mathcal{A}$ are 0 for all but finitely many (s, k). We can find an integer t_0 such that for all s, k we have $c_{s,k} \in c'_{s,k}v^{t_0}1 + v^{t_0-1}\mathbb{Z}[v^{-1}])$ with $c'_{s,k} \in \mathbb{Z}$ for all s, k and $c'_{s,k} \neq 0$ for some s, k.

By 5.24, if $(s, k) \neq (s', k')$, then $\partial(v'^s\mathbf{O}_k, v'^{s'}\mathbf{O}_{k'}) \in v^{-1}\mathbb{Z}[v^{-1}]$. It follows that

$$\partial(\xi', xi') \in 1 + v^{-1}\mathbb{Z}[v^{-1}] = v^{2t_0} = (\sum_{s,k} c'^2_{s,k})v^{2t_0} + v^{2t_0-1}\mathbb{Z}[v^{-1}]$$

Since, from our assumption, we have $\partial(\xi', xi') \in 1 + v^{-1}\mathbb{Z}[v^{-1}]$, it follows that $t_0 = 0$ and $\sum_{s,k} c'^2_{s,k} = 1$. Hence there is a unique value of (s, k) for which $c'_{s,k} = \pm 1$ and $c'_{s,k} = 0$ for all other values of (s, k). In other words,

(a) there is a unique value of (s, k) for which $c_{s,k} \in \pm 1 + v^{-1}\mathbb{Z}[v^{-1}]$ and $c_{s,k} \in v^{-1}\mathbb{Z}[v^{-1}]$ for all other values of (s, k).

Since $\tilde{\beta}(\xi') = \xi'$, we have

$$\sum_{s\in\mathbb{Z},k\in[1,n]} \overline{c_{s,k}}v'^s\mathbf{O}_k = \sum_{s\in\mathbb{Z},k\in[1,n]} c_{s,k}v'^s\mathbf{O}_k$$

where $\overline{}: \mathcal{A} \to \mathcal{A}$ is the ring involution given by $v^t \mapsto v^{-t}$. Hence

(b) $\overline{c_{s,k}} = c_{s,k}$ for all s, k.

Combining (a),(b) we see that there is a unique value of (s, k) for which $c_{s,k} \in \pm 1$ and $c_{s,k} = 0$ for all other values of (s, k). The proposition follows.

5.26

Let S be the algebra of polynomials in two variables x_1, x_2 with coefficients in $\mathbb{C}$. Then $\Gamma = \{r \in \mathbb{C}^* | r^n = 1\}$ acts on S by algebra automorphisms $s : x_1 \mapsto sx_1, s : x_2 \mapsto s^{-1}x_2$. Let $\mathcal{I}$ be the ideal of S generated by the elements x_1x_2, x_1^n, x_2^n. This ideal is Γ-invariant. Let $\tilde{S} = S/\mathcal{I}$. Then $\tilde{S}$ has a basis

$$1, x_1, x_1^2, \ldots, x_1^{n-1}, x_2, x_2^2, \ldots, x_2^{n-1}.$$

On S we have an action of $H = \mathbb{C}^* \times \mathbb{C}^*$ given by $(c', c) : x_1^s x_2^t \mapsto c'^{s-t} c^{s+t} x_1^s x_2^t$. This commutes with the Γ-action.

Each basis element spans a line that is H-stable and Γ-stable. We can number the irreducible characters of Γ as $\rho_k (k \in [0, n-1])$ so that ρ_0 is the unit representation and the ρ_k-isotypic component of $\tilde{S}$ is $\mathbb{C}1$ if $k = 0$ and is $\mathbb{C}x_1^k \oplus \mathbb{C}x_2^{n-k}$ if $k \in [1, n-1]$.

We get an induced H-action on the algebraic variety $\mathbf{H}$ consisting of all ideals $\mathcal{J}$ in S such that $\mathcal{J}$ is Γ-stable and $S/\mathcal{J} \cong Reg_\Gamma$ as a Γ-module ; here, Reg_Γ is the regular representation of Γ. Note that $\mathbf{H}$ is a naturally a closed subvariety of a Hilbert scheme.

Let $\mathbf{H}^0$ be the subvariety of $\mathbf{H}$ considting of all $\mathcal{J} \in \mathbf{H}$ such that $\mathcal{I} \subset \mathcal{J}$. We may identify in an obvious way $\mathbf{H}^0$ with the variety consisting of all ideals of $\tilde{S}$ which are Γ-stable and isomorphic to $Reg_\Gamma - \mathbb{C}$ as a virtual Γ-module.

Assume that $k \in [1, n-1]$. For any $(a, b) \in \mathbb{C}^2 - \{0\}$, the subspace spanned by

$$x_2^{n-1}, x_2^{n-2}, \ldots, x_2^{n-k+1}, ax_1^k + bx_2^{n-k}, x_1^{k+1}, x_1^{k+2}, \ldots, x_1^{n-1} \qquad \text{(a)}$$

is a two sided ideal in $\tilde{S}$ which defines a point of $\mathbf{H}^0$. When a, b vary, we get a subvariety Π_k of $\mathbf{H}^0$ isomorphic to P^1. We have $\cup_k \Pi_k = \mathbf{H}^0$.

For $k \in [1, n]$, let $p_{k-1,k}$ be the subspace spanned by

$$x_2^{n-1}, x_2^{n-2}, \ldots, x_2^{n-k+1}, x_1^k, x_1^{k+1}, x_1^{k+2}, \ldots, x_1^{n-1}.$$

Note that $p_{k-1,k} \in \Pi_k, p_{k,k+1} \in \Pi_k$.

The tangent space to Π_k at $p_{k-1,k}$ is $\text{Hom}(\mathbb{C}x_1^k, \mathbb{C}x_2^{n-k}) = v'^{-n} v^{n-2k}$ as an H-module.

The tangent space to Π_k at $p_{k,k+1}$ is $\text{Hom}(\mathbb{C}x_2^{n-k}, \mathbb{C}x_1^k) = v'^n v^{-n+2k}$ as an H-module.

According to [IN] and [B] there exists an isomorphism $\mathbf{H} \xrightarrow{\sim} \Lambda_e$ which restricts to an isomorphism $\mathbf{H}^0 \xrightarrow{\sim} \mathcal{B}_e$ and to isomorphisms $\Pi_k \xrightarrow{\sim} V_k$ for all $k \in [1, n-1]$. These isomorphisms can be assumed to be compatible with the H-actions. We use them to identify $\mathbf{H} = \Lambda_e$, $\mathbf{H}^0 = \mathcal{B}_e$, $\mathbf{\Pi_k} = \mathbf{V_k}$. Then $p_{k-1,k}$ as just defined is the same as $p_{k-1,k}$ defined in 5.1.

5.27

For $l \in [0, n-1]$ let E^l be the vector bundle over $\mathbf{H}$ whose fibre at $\mathcal{J}$ is $\text{Hom}_\Gamma(\rho_l, S/\mathcal{J})$. The fibre of E^l at the point 5.26(a) is

$\text{Hom}_\Gamma(\rho_l, \mathbb{C}x_1^l)$ if $l < k$,

$\text{Hom}_\Gamma(\rho_l, (\mathbb{C}x_1^k + \mathbb{C}x_2^{n-k})/\mathbb{C}(ax_1^k + bx_2^{n-k}))$ if $k = l$,

$\text{Hom}_\Gamma(\rho_l, \mathbb{C}x_2^{n-l})$ if $l > k$.

Note that $E^0 = \mathbb{C}$. Assume that $l \in [1, n-1]$. If $k \neq l$, the restriction $E^l|_{\Pi_k}$ is a trivial line bundle (if we forget the H-equivariant structure).

The fibre of E^l at $p_{l-1,l}$ is $(\mathbb{C}x_1^l + \mathbb{C}x_2^{n-l})/\mathbb{C}x_1^l = \mathbb{C}x_2^{n-l}$ (hence is $v'^{-n+l}v^{n-l}$).

The fibre of E^l at $p_{l,l+1}$ is $(\mathbb{C}x_1^l + \mathbb{C}x_2^{n-l})/\mathbb{C}x_2^{n-l} = \mathbb{C}x_1^l$ (hence is $v''v^l$).

Hence the restriction $E^l|_{\Pi_l}$ is

$$O_l^{-n+l,n-l,l,l} = v'^{-n+l}O_l^{0,n-l;n,l} = v'^{-n+l}(\mathbf{O}_l)^*.$$

Note also that the fibre of E^l at p_{01} is $\mathbb{C}x_2^{n-l}$ hence it is $v'^{-n+l}v^{n-l}$.

5.28

Let E'^l be the line bundle dual to E^l. Note that $E'^0 = \mathbb{C}$.

Assume that $l \in [1, n-1]$. If $k \neq l$, the restriction $E'^l|_{\Pi_k}$ is a trivial line bundle (if we forget the H-equivariant structure). The restriction $E'^l|_{\Pi_l}$ is $v'^{n-l}\mathbf{O}_l$. The fibre of E'^l at p_{01} is $v'^{n-l}v^{-n+l}$.

Lemma 5.29 Let $l \in [1, n-1]$. We have
(a) $(\mathbf{O}_l\|E'^l) = v'^{-n+l}v^{-2}$;
(b) $(\mathbf{O}_k, E'^l) = 0$ if $k \in [1,n], k \neq l$.

Assume first that $k \in [1, n-1]$. We have

$$(\mathbf{O}_k\|E'^l) = (-v)^{\nu-2}(\varpi^*\tilde{T}_{w_0}\mathbf{O}_k : E'^l)^\dagger$$
$$= -(-v)^{\nu-2}(-1)^\nu v^{-\nu+n}v'^{-n}(\mathbf{O}_k : E'^l)^\dagger$$
$$= -v'^{-n}v^{n-2}(\mathbf{O}_k : E'^l)^\dagger.$$

Now $(\mathbf{O}_k : E'^l)$ is the alternating sum of the cohomologies of Π_k with coefficients in the line bundle $\mathbf{O}_k \otimes E'^l|_{\Pi_k}$. If $l \neq k$, then this is zero. If $l = k$ then $\mathbf{O}_k \otimes E'^k|_{\Pi_k} = v'^{n-k}\mathbf{O}_k \otimes \mathbf{O}_k$. We have

$$\mathbf{O}_k \otimes \mathbf{O}_k = O_k^{0,-n+k;-n,-k}(-O_k^{-n,-k;0,-n+k} + v'^{-n}v^{-k} + v^{-n+k})$$
$$= -O_k^{-n,-n;-n,-n} + (v'^{-n}v^{-k} + v^{-n+k})\mathbf{O}_k$$

and this contributes $-v'^{-n}v^{-n}$ to the Euler characteristic. Hence
(c) $(\mathbf{O}_k : E'^l) = 0$ if $k \neq l$,
$(\mathbf{O}_l : E'^l) = v'^{n-l}(-v'^{-n}v^{-n}) = -v'^{-l}v^{-n}$.
Thus (a) and (b) (with $k \in [1, n-1]$) follow. We have

$$(\mathbf{O}_n\|E''^l) = (-v)^{\nu-2}(\varpi^*\tilde{T}_{w_0}\mathbf{O}_n : E''^l)^\dagger$$

$$= (-v)^{\nu-2}(\varpi^*((-v)^{-\nu}p_{n-1,n} + (-v)^{-\nu}\sum_{k\in[1,n-1]}v^{n-k}\mathbf{O}_k) : E''^l)^\dagger$$

$$= (-v)^{\nu-2}((-v)^{-\nu}p_{01} + (-v)^{-\nu}\sum_{k\in[1,n-1]}v'^n v^{n-k}\mathbf{O}_{n-k} : E''^l)^\dagger$$

$$= (-v)^{-2}(p_{01} + v'^n v^l \mathbf{O}_l : E''^l)^\dagger$$

$$= (-v)^{-2}(v'^{n-l}v^{-n+l} - -v'^n v^j v'^{-j}v^{-n})^\dagger = 0.$$

(We have used 5.19 and (c).) The lemma is proved.

Lemma 5.30 *We have*
(a) $(\mathbf{O}_k\|\mathbb{C}) = 0$ *for all* $k \in [1, n-1]$;
(b) $(\mathbf{O}_n\|\mathbb{C}) = v^{-2}$.

For any $k \in [1, n-1]$ we have
(c) $(\mathbf{O}_k : \mathbb{C}) = 0$.
since the cohomology of P^1 with coefficients in a line bundle with Euler characteristic 0 is 0. We deduce

$$(\mathbf{O}_k\|\mathbb{C}) = (-v)^{\nu-2}(\varpi^*\tilde{T}_{w_0}\mathbf{O}_k : \mathbb{C})^\dagger$$

$$= -(-v)^{\nu-2}(\varpi^*(-1)^\nu v^{-\nu+n}\mathbf{O}_{n-k} : \mathbb{C})^\dagger = -v^{n-2}v'^n(\mathbf{O}_k : \mathbb{C})^\dagger = 0,$$

which proves (a). Using the equalities $(p_{01} : \mathbb{C}) = (p_{n-1,n}, \mathbb{C}) = 1$ and (c) we deduce

$$(\mathbf{O}_n\|\mathbb{C}) = (-v)^{\nu-2}(\varpi^*\tilde{T}_{w_0}\mathbf{O}_n : \mathbb{C})^\dagger$$

$$= (-v)^{\nu-2}(\varpi^*((-v)^{-\nu}p_{01} + (-v)^{-\nu}v'^n\sum_k v^k\mathbf{O}_k) : \mathbb{C})^\dagger$$

$$= v^{-2}(p_{n-1,n} : \mathbb{C})^\dagger = v^{-2}.$$

The lemma is proved.

5.31

For $k \in [1, n]$ we define $\mathbf{e}_k \in Vec_H(\mathbf{H}^0) = Vec_H(\Lambda_e)$ by
$\mathbf{e}_k = v'^{-n+k}v^2 E'^k$ if $k \in [1, n-1]$,
$\mathbf{e}_n = v^2 E'^0 = v^2\mathbb{C}$.
We have
$(\mathbf{O}_{k'}\|\mathbf{e}_k) = \delta_{\mathbf{k},\mathbf{k}'}$ for all $k, k' \in [1, n]$.
In particular, we see that the pairing $(\|)$ is non-singular and that $\{\mathbf{e}_k | k \in [1, n]\}$ is an R_H-basis of $K_H(\Lambda_e)$. Arguing now as in [[L6], 7.9], we see that $\pm v'^s\mathbf{O}_k(k \in [1, n], s \in \mathbb{Z})$ coincides with the set $\mathbf{B}^\pm_{\Lambda_e}$ defined in [[L5], 5.11] (a signed basis of $K_H(\Lambda_e)$).

6 Subregular case: types C, D, E, F, G

6.1

In this section we assume that G is of type C, D, E, F or G. In this case we have $C = \{1\}$ hence $H = \mathbb{C}^*$. We want to describe the set $\mathbf{B}^{\pm}_{\mathcal{B}_e}$. The case where G is of type D or E is discussed in [L6]; here we shall review the results in [L6] including at the same time G of type C, F, G. We have
$$J = \{j_0^1, j_1^1, \ldots, j_{a_1}^1\} \cup \{j_0^2, j_1^2, \ldots, j_{a_2}^2\} \cup \{j_0^3, j_1^3, \ldots, j_{a_3}^3\}$$
(a disjoint union except for $j_0^1 = j_0^2 = j_0^3$) where a_1, a_2, a_3 are ≥ 1,
   j, j' are joined in J precisely when $\{j, j'\} = \{j_t^u, j_{t+1}^u\}$ with $u \in \{1, 2, 3\}$, $0 \leq t < a_u$.
We denote $j_0^1 = j_0^2 = j_0^3$ by j_0.
   For $u \in \{1, 2, 3\}$, $0 \leq t < a_u$, we write $p_{t,t+1}^u$ instead of $p_{j_t^u, j_{t+1}^u}$. Note that $p_{0,1}^1, p_{0,1}^2, p_{0,1}^3$ are distinct points of V_{j_0}, but they are the same as elements of $K_{\mathbb{C}^*}$ or $K_{\mathbb{C}^*}(\mathcal{B}_e)$ which are denoted by p.
   The fixed point set of the $H = \mathbb{C}^*$-action of $\mathcal{B}_e$ has connected components μ_j $(j \in J)$ where
   $\mu_j = V_j$ if $j = j_0$;
   $\mu_j = \{p_{t,t+1}^u\}$ if $j = j_t^u$ with $u \in \{1, 2, 3\}$, $0 < t < a_u$,
   $\mu_j = \{q^u\}$ if $j = j_t^u$ with $u \in \{1, 2, 3\}$, $t = a_u$.
Thus, if $j = j_t^u$ with $u \in \{1, 2, 3\}$, $t = a_u$, there are two H-fixed points on V_j, namely $p_{t-1,t}^u$ and q^u.

6.2

There is a unique homomorphism $n_0 : X \to \mathbb{Z}$ such that
   $n_0(\alpha_i) = -2$ if $i \neq \omega(j_0)$, $n_0(\alpha_i) = 0$ if $i = \omega(j_0)$.
For $j \in J$ we define a homomorphism $n_j : X \to \mathbb{Z}$ by $n_{j_0} = n_0$ and
   $n_i(x) = n_0(s_{i_1^u} s_{i_2^u} \ldots s_{i_t^u}(x))$
where $j = j_t^u, u \in \{1, 2, 3\}, 0 < t \leq a_u$ and $i_r^u = \omega(j_r^u)$ for $r \in [1, t]$.
   If $x \in X$, we regard $L_x \in Vec_{\mathcal{G}}(\mathcal{B})$ as an object of $Vec_{\mathbb{C}^*}(\mathcal{B})$ by restriction via $\mathbb{C}^* = H \subset \mathcal{G}$. In particular we obtain a $\mathbb{C}^*$-action on the fibre of L_x at a $\mathbb{C}^*$-fixed point on $\mathcal{B}_e$.
   (a) Let $j \in J, x \in X$ and let $\mathfrak{b} \in \mu_j$. Then $\mathbb{C}^*$ acts on the fibre of L_x through the character $v^{n_j(x)}$.
This is proved as in [[L6], 3.2].

6.3

Fix $j \in J$ and $m \in \mathbb{Z}$. There is, up to isomorphism, a unique $\mathbb{C}^*$-equivariant line bundle O_j^m on V_j such that
   (a) the Euler characteristic of V_j with coefficients in O_j^m is $m + 1$;
   (b) if $j = j_t^u$, then $\mathbb{C}^*$ acts on the fibre of O_j^m at a point of μ_j by v^{tm}.
If $j = j_0$ and $f : \{p_{0,1}^u\} \to V_j$ is the inclusion, we have

$$p = O_j^0 - O_j^{-1} \in K_{\mathbb{C}^*}(V_j).$$

6.4

We regard O_j^m as an object of $Coh_{\mathbb{C}^*}(\mathcal{B}_e)$ (zero outside V_j); this object is denoted again by O_j^m. As in [L6], we see that $K_{\mathbb{C}^*}(\mathcal{B}_e)$ is the $\mathcal{A}$-module with generators O_j^m ($j \in J, m \in \mathbb{Z}$) and relations

$$O_{j_t^u}^0 - v^{-t}O_{j_t^u}^{-1} = O_{j_{t+1}^u}^0 - v^{t+1}O_{j_{t+1}^u}^{-1}$$

for $u \in \{1, 2, 3\}, 0 \le t < a_u,$

$$O_j^{m+1} + O_j^{m-1} = (v^t + v^{-t})O_j^m$$

for $j = j_t^u, 0 \le t \le a_u, m \in \mathbb{Z}.$

It follows that

(a) an $\mathcal{A}$-basis of $K_{\mathbb{C}^*}(\mathcal{B}_e)$ is given by O_j^{-1}($j \in J$) and $p = O_{j_0}^0 - O_{j_0}^{-1}.$

6.5

For $x \in X$, the restriction of L_x to V_j is $v^s O_j^{\check{\alpha}_i(x)}$ where $i = \omega(j)$ and $s = n_j(x) - t\check{\alpha}_i(x)$ (with $j = j_t^u$).

6.6

As in [[L6], 3.6], we have

(a) $\theta_x p = v^{n_0(x)}p;$

(b) if $j = j_t^u, i = \omega(j)$ and $\check{\alpha}_i(x) = 1$, then $\theta_x O_j^m = v^{n_j(x)-t}O_j^{m+1}$ and $\theta_{x-\alpha_i}O_j^m = v^{n_j(x)-t}O_j^{m-1}.$

Lemma 6.7 Let $j = j_t^u, j' = j_{t-1}^u$ with $u \in \{1, 2, 3\}, 0 < t \le a_u$. Let $\tilde{p} = p_{t-1,t}^u$. Let $i = \omega(j), i' = \omega(j')$. We have

(a) $\tilde{T}_i\tilde{p} = -v^{-1}\tilde{p} + (v^{t-1} - v^{-t+1})O_j^{-1};$

(b) $\tilde{T}_{i'}\tilde{p} = -v^{-1}\tilde{p} + (v^t - v^{-t})O_{j'}^{-1}.$

The proof (based on 4.7(b) and 6.6) is along the same lines as that of [[L6], 3.9]. (The argument in [[L6], 3.9] can be simplified since here we know in advance that the coefficient of $\tilde{p}$ in (a) and (b) is $-v^{-1}$ (by 4.7(b)).

6.8 L

et $i_0 = \omega(j_0)$. The following is a special case of 6.7:

(a) $\tilde{T}_{i_0}p = -v^{-1}p + (v - v^{-1})O_{j_0}^{-1}.$

Lemma 6.9 Let $i \in I$ be such that $i \ne \omega(j_0)$. Then

(a) $\tilde{T}_i p = -v^{-1}p.$

This follows from 4.7(e) except in the special case where $\cup_{j;\omega(j)=i}V_j$ contains all three points p_{01}^u. In this special case, G is be of type G_2 and i is uniquely determined. For $j \neq j_0$, V_j is i-saturated and contains one of the points p_{01}^u hence $\tilde{T}_i p = -v^{-1}p + a_j O_j^{-1}$ where $a_j \in \mathcal{A}$. (We use 4.7(c).) Thus $a_j O_j^{-1}$ takes the same value for three different j. It follows that $a_j = 0$ and (a) is proved.

Lemma 6.10 *If $i \in I, j \in J$ are such that $i = \omega(j)$, then*
   (a) $\tilde{T}_i(O_j^{-1}) = vO_j^{-1}$.

This is a special case of 4.7(a).

Lemma 6.11 *Assume that $i \in I, j \in J$ are such that $i \neq \omega(j)$. Then*
   (a) $\tilde{T}_i(O_j^{-1}) = -v^{-1}O_j^{-1} - \sum_{j' \in J'} O_{j'}^{-1}$
where J' consists of all $j' \in J$ that are joined with j and satisfy $\omega(j') = i$.

If $J' = \emptyset$, then (a) follows from 4.7(d). Assume now that $j = j_t^u, j' = j_{t-1}^u$ with $0 < t \leq a_u$ and $i = \omega(j')$. Then (a) is proved in the same way as [[L6], 3.11(a)] (using 4.7(d) and 6.7); again the proof in [[L6], 3.11] is simplified by the use of 4.7. Assume next that $j = j_{t-1}^u, j' = j_t^u$ with $1 < t \leq a_u$ and $i = \omega(j')$. Then (a) is proved in the same way as [[L6], 3.11(b)].

It remains to consider the case where $j = j_0, i = \omega(j_1^u)$. We use 4.7(d). By symmetry, the coefficients $c_{j'}$ in 4.7(d) are independent of j'. Thus,
$$\tilde{T}_i(O_{j_0}^{-1}) = -v^{-1}O_{j_0}^{-1} + c\sum_{j' \in J'} O_{j'}^{-1} \text{ with } c \in \mathcal{A}.$$
Setting $f = \sum_{j' \in J'} O_{j'}^{-1}$, we have
   (b) $\tilde{T}_i(O_{j_0}^{-1}) = -v^{-1}O_{j_0}^{-1} + cf$.
From the earlier part of the proof we deduce
   (c) $\tilde{T}_{i_0}f = -v^{-1}f - |J'|O_{j_0}^{-1}$
where $i_0 = \omega(j_0)$. Using (b),(c) we can express the braid group relation
   $\ldots \tilde{T}_{i_0}\tilde{T}_i\tilde{T}_{i_0} = \ldots \tilde{T}_i\tilde{T}_{i_0}\tilde{T}_i$
on the $\mathcal{A}$-submodule spanned by f, O_{j_0} as an equality of two explicit 2×2 matrices with entries in $\mathcal{A}$. This equality gives us $c = -1$ if $|J'|$ is 1 or 2 and $(c+1)(3c+1) = 0$ if $|J'| = 3$. In the last case we must also have $c = -1$ since $c \in \mathcal{A}$. This proves (a).

Lemma 6.12 *Let $j \mapsto j^*$ be the opposition involution of the Coxeter graph J. Let $\mathbf{h}$ be the Coxeter number of G. This is the same as the Coxeter number of the graph J. We have*
   (a) $\tilde{T}_{w_0}(O_j^{-1}) = -(-v)^{-\nu+\mathbf{h}}O_{j^*}^{-1}$;
   (b) $\tilde{T}_{w_0}^{-1}(O_j^{-1}) = -(-v)^{\nu-\mathbf{h}}O_{j^*}^{-1}$.

When G is of type D, E, this is proved in [[L6], 4.5]. Assume now that G is of type $C_n(n \geq 3), F_4$ or G_2. Let σ be a generator of the cyclic group $\mathbb{Z}/d\mathbb{Z}$ which acts on J ($d \in \{2,3\}$) as in 4.2.

Let $\mathcal{H}_0$ be the $\mathcal{A}$-subalgebra of $\mathcal{H}$ generated by $\{\tilde{T}_i | i \in I\}$. Let $\mathcal{M}$ be the $\mathcal{A}$-submodule of $K_H(\mathcal{B}_e)$ with basis $\{O_j^{-1} | j \in J\}$. Then $\mathcal{M}$ is an $\mathcal{H}_0$-submodule of $K_H(\mathcal{B}_e)$.

Let $\mathcal{M}'$ be the $\mathcal{A}$-submodule of $\mathcal{M}$ generated by the elements O_j^{-1} with $\sigma(j) = j$ and $O_j^{-1} + O_{\sigma(j)}^{-1}$ with $\sigma(j) \neq j$; let $\mathcal{M}''$ be the $\mathcal{A}$-submodule of $\mathcal{M}$ generated by the elements $O_j^{-1} - O_{\sigma(j)}^{-1}, O_j^{-1} - O_{\sigma^{-1}(j)}^{-1}$ with $\sigma(j) \neq j$. Then $\mathcal{M}', \mathcal{M}''$ are $\mathcal{H}$-submodules and their direct sum is $\mathcal{M}$ (after extending scalars to $\mathbb{Q}(v)$). Now $\mathcal{M}'$ is irreducible as an $\mathcal{H}$-module (if scalars are extended to an algebraic closure of $\mathbb{Q}(v)$); the same holds for $\mathcal{M}''$ except for type G_2. Now $\tilde{T}_{w_0}$ is central in $\mathcal{H}_0$ (since in our case w_0 is central in W.) Hence $\tilde{T}_{w_0}$ acts on $\mathcal{M}'$ as multiplication by $b' \in \mathcal{A}$. Similarly, (in type C, F), $\tilde{T}_{w_0}$ acts on $\mathcal{M}''$ as multiplication by $b'' \in \mathcal{A}$.

Now b' can be determined as in [[L6], 4.5] since for $v = 1$, $\mathcal{M}'$ becomes the reflection representation of W. We see that $b' = -(-v)^{-\nu+\mathbf{h}}$.

For G of type C_n, $\mathcal{M}''$ has rank 1 and any $\tilde{T}_i$ such that $|\omega^{-1}(i)| = 1$ acts as $-v^{-1}$ on it, while the $\tilde{T}_i$ such that $\omega^{-1}(i) = 2$ acts as v on it. Hence $\tilde{T}_{w_0}$ acts on it as $(-v^{-1})^{n^2-n}v^n = v^{-\nu+\mathbf{h}}$.

For G of type F_4, $\mathcal{M}''$ has rank 2 and any $\tilde{T}_i$ such that $|\omega^{-1}(i)| = 1$ acts as $-v^{-1}$ on it, while any $\tilde{T}_i$ such that $|\omega^{-1}(i)| = 1$ acts with eigenvalues $-v^{-1}, v$. Hence the determinant of $\tilde{T}_{w_0}$ on $\mathcal{M}''$ is v^{-24}. Hence $\tilde{T}_{w_0}$ acts on $\mathcal{M}''$ as $\pm v^{-12}$. If we specialize v to 1, w_0 acts as 1. Hence $\tilde{T}_{w_0}$ acts on $\mathcal{M}''$ as v^{-12}.

For G of type G_2, $\mathcal{M}''$ has rank 2 and one checks easily that $\tilde{T}_{w_0}$ acts on it as -1.

Thus, the action on $\tilde{T}_{w_0}$ on $\mathcal{M}', \mathcal{M}''$ is explicitly described, hence its action on $\mathcal{M}$ is determined and (a) follows. Now (b) follows from (a).

6.13

We consider the system of equations with unknowns $A_j (j \in J)$:

$(v + v^{-1})A_j = \sum_{j' \in J; j, j' \text{ joined}} A_{j'}$, if $j \in J - \{j_0\}$,

$(v + v^{-1})A_j = \sum_{j' \in J; j, j' \text{ joined}} A_{j'} - (v - v^{-1})$, if $j = j_0$.

This system has a unique solution with $A_j \in \mathbb{Q}(v)$ for all $j \in J$. It satisfies $A_j = \frac{\bar{B}_j - B_j}{v^{\mathbf{h}/2} - v^{-\mathbf{h}/2}}$ where $B_j \in v\mathbb{Z}[v]$ for all $j \in J$ and $\bar{\ }: \mathcal{A} \to \mathcal{A}$ is as in 5.25. (See [[L6], 4.1, 4.2].) Note that $\mathbf{h}$ is even in our case. The polynomials B_j are given explicitly in [[L6], 1.10].

Lemma 6.14 We have (a) $\tilde{T}_{w_0}^{-1}(p) = (-v)^{\nu}p + (-v)^{\nu}(1 + v^{-\mathbf{h}})\sum_{j \in J} A_j O_j^{-1}$;

(b) $\tilde{T}_{w_0}(p) = (-v)^{-\nu}p + (-v)^{-\nu}(1 + v^{\mathbf{h}})\sum_{j \in J} A_j O_j^{-1}$.

Let $\xi = p + \sum_{j \in J} A_j O_j^{-1}$. Using 6.8-6.11, the definition of A_j and the fact that A_j is constant on any $\mathbb{Z}/d\mathbb{Z}$ orbit in J, we see that $\tilde{T}_i \xi = -v^{-1}\xi$ for all $i \in I$. It follows that $\tilde{T}_{w_0}\xi = (-v)^{-\nu}\xi$ or equivalently

$\tilde{T}_{w_0}(p + \sum_{j \in J} A_j O_j^{-1}) = (-v)^{-\nu}(p + \sum_{j \in J} A_j O_j^{-1})$.

Using 6.12 and the equality $A_{j^*} = A_j$ we deduce

$\tilde{T}_{w_0}(p) - \sum_{j \in J} A_j(-v)^{-\nu+\mathbf{h}}O_j^{-1} = (-v)^{-\nu}(p + \sum_{j \in J} A_j O_j^{-1})$

and (b) follows. Next note that $\tilde{T}_{w_0}^{-1}\xi = (-v)^{\nu}\xi$ or equivalently

$$\tilde{T}_{w_0}^{-1}(p + \sum_{j \in J} A_j O_j^{-1}) = (-v)^\nu (p + \sum_{j \in J} A_j O_j^{-1}).$$

Using again 6.12 we deduce

$$\tilde{T}_{w_0}^{-1}(p) - \sum_{j \in J} A_j (-v)^{\nu - \mathbf{h}} O_j^{-1} = (-v)^\nu (p + \sum_{j \in J} A_j O_j^{-1})$$

and (a) follows.

Lemma 6.15 *Let* $\mathbf{p} = p - \sum_{j \in J} B_j v^{-\mathbf{h}/2} O_j^{-1}$. *We have*

$$\tilde{T}_{w_0}(\mathbf{p}) = (-v)^{-\nu} (p + \sum_{j \in J} v^{\mathbf{h}/2} B_j O_j^{-1}).$$

This follows immediately from 6.12, 6.13, 6.14. (Compare [[L6], 4.7].)

6.16

Let $\underline{G}$ be a connected semisimple simply connected algebraic group over $\mathbb{C}$ whose Coxeter graph is J as in 2.2. The analogues for $\underline{G}$ of the various objects attached to G will be denoted by the same symbol underlined. Thus,

$$\underline{I}, \underline{\mathfrak{g}}, \underline{W}, \underline{w_0}, \underline{\nu}, \underline{\mathbf{h}}, \underline{\mathcal{H}}, \underline{\tilde{T}_{w_0}} \in \underline{\mathcal{H}}$$

are defined.

We fix an $\mathfrak{sl}_2$-triple $\underline{e}, \underline{h}, \underline{f}$ in $\underline{\mathfrak{g}}$ such that $\underline{e}$ is subregular in $\underline{\mathfrak{g}}$. Let $\mathcal{B}_{\underline{e}}, \Lambda_{\underline{e}}, \underline{\varpi}, \underline{J}$ be the analogues for $\underline{G}$ of $\mathcal{B}_e, \Lambda_e, \varpi, J$. Let $(\underline{:}), (\underline{||}), \underline{\tilde{\beta}}, \underline{k}$ be the analogues for $\underline{G}$ of $(:), (||), \tilde{\beta}, k$ (k as in 4.5).

Note that $\underline{I} = J$ and $\underline{\mathbf{h}} = \mathbf{h}$. Moreover, according to [S] we can find a $\mathbb{C}^*$-equivariant isomorphism $\Lambda_{\underline{e}} \xrightarrow{\sim} \Lambda_e$ which carries $\mathcal{B}_{\underline{e}}$ onto $\mathcal{B}_e$ and is such that the induced map $\underline{J} \to J$ is the identity map. (We have $\underline{J} = \underline{I} = J$.) Under this isomorphism $\underline{k}$ corresponds to k. Using these isomorphisms we may identify $K_{\mathbb{C}^*}(\mathcal{B}_{\underline{e}}) = K_{\mathbb{C}^*}(\mathcal{B}_e)$ and $K_{\mathbb{C}^*}(\Lambda_{\underline{e}}) = K_{\mathbb{C}^*}(\Lambda_e)$. It is then clear that $(\underline{:}) = (:)$ and $D_{\mathcal{B}_e} = D_{\mathcal{B}_{\underline{e}}}$. It is also clear that, for $j \in J = \underline{J}$, the element O_j^{-1} defined in terms of G is the same as that defined in terms of $\underline{G}$. From the definition, we have $\varpi^* O_j^{-1} = O_{j^*}^{-1}$, $\underline{\varpi}^* O_j^{-1} = O_j^{-1}$. Also, p defined in terms of G is the same as p defined in terms of $\underline{G}$. Moreover, $\varpi^*(p) = p, \underline{\varpi}^*(p) = p$. Since the elements O_j^{-1} together with p form an $\mathcal{A}$-basis of $K_{\mathbb{C}^*}(\mathcal{B}_e)$, it follows that $\varpi^* = \underline{\varpi}^*$ on $K_{\mathbb{C}^*}(\mathcal{B}_e)$.

Comparing 6.12, 6.14 with the analogous formulas for $\underline{G}$ we see that

$$\tilde{T}_{\underline{w_0}} = (-v)^{-\underline{\nu} + \nu} \tilde{T}_{w_0} : K_{\mathbb{C}^*}(\mathcal{B}_e) \to K_{\mathbb{C}^*}(\mathcal{B}_e).$$

Using this we see that

$$(F \underline{||} F') = (-v)^{\underline{\nu} - 2} (\underline{\varpi}^* \tilde{T}_{\underline{w_0}} F : F') = (-v)^{\nu - 2} (\varpi^* \tilde{T}_{w_0} F : F') = (F || F'),$$

$$\underline{\tilde{\beta}} = (-v)^{-\underline{\nu}} \underline{\varpi}^* \tilde{T}_{\underline{w_0}}^{-1} D_{\mathcal{B}_e} = (-v)^{-\nu} \varpi^* \tilde{T}_{w_0}^{-1} D_{\mathcal{B}_e} = \tilde{\beta}.$$

It follows that $\mathbf{B}_{\mathcal{B}_e}^\pm$ defined in terms of G is the same as $\mathbf{B}_{\mathcal{B}_e}^\pm$ defined in terms of $\underline{G}$. Since the latter is known from [[L6], 6.4], we see that the following holds (for G):

Proposition 6.17 $\mathbf{B}^{\pm}_{\mathcal{B}_e}$ is the signed $\mathcal{A}$-basis of $K_{\mathbb{C}^*}(\mathcal{B}_e)$ consisting of $\pm$ the elements $v^{-h/2}O_j^{-1}(j \in J)$ and $\mathbf{p}$.

Here $\mathbf{p}$ is define in terms of $\underline{G}$.

Proposition 6.18 There exist $\mathbb{C}^*$-equivariant vector bundles $E'^j (j \in J)$ on Λ_e such that

(a) $v^2 E'^{j*} (j \in J), v^2 \mathbb{C}$ form an $\mathcal{A}$-basis of $K_{\mathbb{C}^*}(\Lambda_e)$ dual to the basis $v^{-h/2}O_j^{-1}(j \in J), \mathbf{p}$ of $K_{\mathbb{C}^*}(\mathcal{B}_e)$ with respect to $(||)$.

(b) $\mathbf{B}^{\pm}_{\Lambda_e}$ (see $[[L5], 5.11]$) is the signed $\mathcal{A}$-basis of $K_{\mathbb{C}^*}(\Lambda_e)$ consisting of $\pm$ the elements $v^2 E'^{j*} (j \in J), v^2 \mathbb{C}$.

The arguments above reduce this to the case where G is replaced by $\underline{G}$, which is known by $[[L6], 7.7, 7.9]$.

6.19

Let $\mathfrak{z}$ be the centralizer of f in $\mathfrak{g}$. This is an H-stable subspace of $\mathfrak{g}$. The element $\sum_{t \geq 0}(-1)^t \mathfrak{z}^{(t)} \in R_H = \mathcal{A}$ (where $\mathfrak{z}^{(t)}$ is the t-th exterior power of $\mathfrak{z}$) is divisible in $\mathcal{A}$ by $(1 - v^{-2})^r$ (where r is the rank of $\mathfrak{g}$); see $[[L5], 3.2]$. The quotient is denoted by $\nabla_e \in \mathcal{A}$. If $\xi \in K_{\mathbb{C}^*}(\Lambda_e)$, there is a unique element $m(\xi) \in K_{\mathcal{B}^*}(\mathcal{B}_e)$ such that $k_*(m(\xi)) = \nabla_e \xi$ with k as in 4.5 (see $[[L5], 3.5]$).

Let $\tilde{J} = J \sqcup \{0\}$. For $j \in \tilde{J}$ we define $b_j \in K_{\mathbb{C}^*}(\Lambda_e)$ by $b_j = v^2 E'^{j*}$ if $j \in J$, $b_0 = v^2 \mathbb{C}$. We want to compute the inner products $(m(b_j)||b_{j'}) \in \mathcal{A}$ for any $j, j' \in \tilde{J}$. (These products are of interest in connection with the representation theory of Lie algebras in characteristic p; see [L7].)

Let $\mathcal{P} = \sum_{w \in W} v^{-2l(w)} \in \mathcal{A}$.

Let $\underline{m} : K_{\mathbb{C}^*}(\underline{\Lambda_e}) \to K_{\mathbb{C}^*}(\underline{\mathcal{B}_e}), \underline{\mathcal{P}}, \underline{\nabla_e}$ be the analogues for $\underline{G}$ of $m : K_{\mathbb{C}^*}(\Lambda_e) \to K_{\mathbb{C}^*}(\mathcal{B}_e), \mathcal{P}, \nabla_e$.

The analogue of b_j for $\underline{G}$ is b_j itself (see 6.17, 6.18).

Proposition 6.20 $(m(b_j)||b_{j'}) = \frac{\mathcal{P}}{\underline{\mathcal{P}}}(\underline{m}(b_j)||b_{j'})$.

(Note that $(\underline{m}(b_j)||b_{j'})$ is explicitly computed in [L7].)

From the definitions we have $(m(b_j)||b_{j'}) = \frac{\nabla_e}{\underline{\nabla_e}}(\underline{m}(b_j)||b_{j'})$. It remains to observe that

(a) $\frac{\nabla_e}{\underline{\nabla_e}} = \frac{\mathcal{P}}{\underline{\mathcal{P}}}$.

Indeed, by computation, we see that both sides of (a) are equal to

$(1 - v^{-2})(1 - v^{-2(n+1)})^{-1}$ (for type C_n),
$(1 - v^{-2})^2(1 - v^{-10})^{-1}(1 - v^{-18})^{-1}$ (for type F_4),
$(1 - v^{-2})^2(1 - v^{-8})^{-2}$ (for type G_2);

for types D, E both sides of (a) are clearly 1.

References

[B]   E. Brieskorn: *Singular elements of semisimple algebraic groups*, Actes Congrès Intern. Math., (1970), textbf2, 279–284

[DLP]   C. De Concini, G. Lusztig and C. Procesi: *Homology of the zero-set of a nilpotent vector field on a flag manifold*, J. Amer. Math.Soc. ,1, 1988 15–34

[GI]   V. Ginzburg: *Lagrangian construction of representations of Hecke algebras*, Adv. in Math., **63**, (1987), 100–112

[IM]   N. Iwahori and H. Matsumoto: *On some Bruhat decompositions and the structure of the Hecke rings of p-adic Chevalley groups*,Inst. Hautes Études Sci. Publ. Math.,**25**, (1965) 5–48

[IN]   Y. Ito and I. Nakamura: *McKay correspondence and Hilbert schemes* ,Proc. Japan Acad. Sci. A , **72**, (1996) 135–138

[KL]   D. Kazhdan and G. Lusztig: *Proof of the Deligne-Langlands conjecture for Hecke algebras*, Inv. Math., **87**, (1987) 153–215

[KT]   M. Kashiwara and T. Tanisaki: *The characteristic cycles of holonomic systems on a flag manifold*, Inv. Math., **77**,(1984) 185–198

[L1]   G. Lusztig: *Some examples of square integrable representations of semisimple p-adic groups*, Trans. Amer. Math. Soc., **277**, (1983) 623–653

[L2]   G. Lusztig: *Equivariant K-theory and representations of Hecke algebras*, Proc. Amer. Math. Soc., **94**,(1985) 337–342

[L3]   G. Lusztig: *Affine Hecke algebras and their graded version*, J. Amer. Math. Soc. ,**2**, (1989) 599–635

[L4]   G. Lusztig: *Bases in equivariant K-theory* Represent. Th. (electronic), **2**, (1998) 298–369

[L5]   G. Lusztig: *Bases in equivariant K-theory, II*, Represent. Th. (electronic),**3**, (1999), 281–353

[L6]   G. Lusztig: *Subregular nilpotent elements and bases in K-theory*, Canad. J. Math., **51**, (1999), 1194–1225

[L7]   G. Lusztig: *Representation theory in characteristic p*, Taniguchi Conf. on Math. Nara'98", Adv.Stud.Pure Math., **31**, Math.Soc.Japan, (2001), 167–178

[S]   P. Slodowy: *Simple algebraic groups and simple singularities*, Lecture Notes in Math. 815, (1980), Springer Verlag Berlin-Heidelberg-New York

List of Participants

1. N.B. Andersen Denmark Paris
2. A. Artemov Russia Tambov
3. Y. Bazlov Russia Israel
4. F. Bernon France Poitiers
5. G. Carnovale Italy Utrecht
6. N. Ciccoli Italy Perugia
7. I. Damiani Italy Roma
8. A. D'Andrea Italy Strasbourg
9. F. Gavarini Italy Roma
10. V. Guizzi Italy Roma
11. T. Haines USA Canada
12. V. Heierman Germany Berlin
13. J. Kim Korea USA
14. B. Klingler France Paris
15. C. Kriloff U.S.A. Idaho
16. G. Kuhn Italy Milan
17. U. Kulkarni India USA
18. B. Lemaire France Paris
19. A. Maffei Italy Roma
20. K. Maktouf Tunisia Poitiers
21. K. McGerty Irish Boston
22. Y. Markov Bulgaria USA
23. P. Moseneder Italy Trento
24. L. Migliorini Italy Firenze
25. M. Nazarov Russia York
26. G. Papadopoulo France Roma
27. V. Protsak Ukraine USA
28. J.F. Quint France Paris
29. D. Renard France Poitiers
30. K. Reitsch Austria Cambridge
31. A. Sarveniazi Iran Goettingen
32. K. Slooten Netherlands Leiden
33. E. Sommers USA USA
34. K. Styrkas Russia USA
35. C. Urtis Turkey USA

LIST OF C.I.M.E. SEMINARS

1966	39. Calculus of variations	Ed. Cremonese, Firenze
	40. Economia matematica	"
	41. Classi caratteristiche e questioni connesse	"
	42. Some aspects of diffusion theory	
1967	43. Modern questions of celestial mechanics	"
	44. Numerical analysis of partial differential equations	"
	45. Geometry of homogeneous bounded domains	"
1968	46. Controllability and observability	"
	47. Pseudo-differential operators	"
	48. Aspects of mathematical logic	"
1969	49. Potential theory	"
	50. Non-linear continuum theories in mechanics and physics and their applications	"
	51. Questions of algebraic varieties	"
1970	52. Relativistic fluid dynamics	"
	53. Theory of group representations and Fourier analysis	"
	54. Functional equations and inequalities	"
	55. Problems in non-linear analysis	"
1971	56. Stereodynamics	"
	57. Constructive aspects of functional analysis (2 vol.)	"
	58. Categories and commutative algebra	"
1972	59. Non-linear mechanics	"
	60. Finite geometric structures and their applications	"
	61. Geometric measure theory and minimal surfaces	"
1973	62. Complex analysis	"
	63. New variational techniques in mathematical physics	"
	64. Spectral analysis	"
1974	65. Stability problems	"
	66. Singularities of analytic spaces	"
	67. Eigenvalues of non linear problems	"
1975	68. Theoretical computer sciences	"
	69. Model theory and applications	"
	70. Differential operators and manifolds	"
1976	71. Statistical Mechanics	Ed. Liguori, Napoli
	72. Hyperbolicity	"
	73. Differential topology	"
1977	74. Materials with memory	"
	75. Pseudodifferential operators with applications	"
	76. Algebraic surfaces	"
1978	77. Stochastic differential equations	Ed. Liguori, Napoli
	78. Dynamical systems	&
		Birkhäuser
1979	79. Recursion theory and computational complexity	"
	80. Mathematics of biology	"

1980	81. Wave propagation		Ed. Liguori, Napoli
	82. Harmonic analysis and group representations		&
	83. Matroid theory and its applications		Birkhäuser
1981	84. Kinetic Theories and the Boltzmann Equation	(LNM 1048)	Springer-Verlag
	85. Algebraic Threefolds	(LNM 947)	"
	86. Nonlinear Filtering and Stochastic Control	(LNM 972)	"
1982	87. Invariant Theory (LNM 996)		"
	88. Thermodynamics and Constitutive Equations	(LN Physics 228)	"
	89. Fluid Dynamics	(LNM 1047)	"
1983	90. Complete Intersections	(LNM 1092)	"
	91. Bifurcation Theory and Applications	(LNM 1057)	"
	92. Numerical Methods in Fluid Dynamics	(LNM 1127)	"
1984	93. Harmonic Mappings and Minimal Immersions	(LNM 1161)	"
	94. Schrödinger Operators	(LNM 1159)	"
	95. Buildings and the Geometry of Diagrams	(LNM 1181)	"
1985	96. Probability and Analysis	(LNM 1206)	"
	97. Some Problems in Nonlinear Diffusion	(LNM 1224)	"
	98. Theory of Moduli	(LNM 1337)	"
1986	99. Inverse Problems	(LNM 1225)	"
	100. Mathematical Economics	(LNM 1330)	"
	101. Combinatorial Optimization	(LNM 1403)	"
1987	102. Relativistic Fluid Dynamics	(LNM 1385)	"
	103. Topics in Calculus of Variations	(LNM 1365)	"
1988	104. Logic and Computer Science	(LNM 1429)	"
	105. Global Geometry and Mathematical Physics	(LNM 1451)	"
1989	106. Methods of nonconvex analysis	(LNM 1446)	"
	107. Microlocal Analysis and Applications	(LNM 1495)	"
1990	108. Geometric Topology: Recent Developments	(LNM 1504)	"
	109. H_∞ Control Theory	(LNM 1496)	"
	110. Mathematical Modelling of Industrial Processes	(LNM 1521)	"
1991	111. Topological Methods for Ordinary Differential Equations	(LNM 1537)	"
	112. Arithmetic Algebraic Geometry	(LNM 1553)	"
	113. Transition to Chaos in Classical and Quantum Mechanics	(LNM 1589)	"
1992	114. Dirichlet Forms	(LNM 1563)	"
	115. D-Modules, Representation Theory, and Quantum Groups	(LNM 1565)	"
	116. Nonequilibrium Problems in Many-Particle Systems	(LNM 1551)	"

Fondazione C.I.M.E.

Centro Internazionale Matematico Estivo
International Mathematical Summer Center
http://www.math.unifi.it/∼cime
cime@math.unifi.it

2003 COURSES LIST

Stochastic Methods in Finance

July 6–13, Cusanus Akademie, Bressanone (Bolzano)
Joint course with European Mathematical Society

Course Directors:

Prof. Marco Frittelli (Univ. di Firenze), marco.frittelli@dmd.unifi.it
Prof. Wolfgang Runggaldier (Univ. di Padova), runggal@math.unipd.it

Hyperbolic Systems of Balance Laws

July 14–21, Cetraro (Cosenza)

Course Director:

Prof. Pierangelo Marcati (Univ. de L'Aquila), marcati@univaq.it

Symplectic 4-Manifolds and Algebraic Surfaces

September 2–10, Cetraro (Cosenza)

Course Directors:

Prof. Fabrizio Catanese (Bayreuth University)
Prof. Gang Tian (M.I.T. Boston)

Mathematical Foundation of Turbulent Viscous Flows

September 1–6, Martina Franca (Taranto)

Course Directors:

Prof. M. Cannone (Univ. de Marne-la-Vallée)
Prof.T. Miyakawa (Kobe University)

Printing and Binding: Strauss GmbH, Mörlenbach